GUIDE TO MATHEMATICAL MODELLING

MACMILLAN MATHEMATICAL GUIDES

Consultant Editor: **David A. Towers,**
Senior Lecturer in Mathematics,
University of Lancaster

Linear Algebra D. Towers
Abstract Algebra C. Whitehead
Analysis F. Hart
Numerical Analysis J. Turner
Mathematical Modelling D. Edwards and M. Hanson
Mathematical Methods J. Gilbert
Mechanics P. Dyke and R. Whitworth

Guide to Mathematical Modelling

Dilwyn Edwards and Mike Hamson

School of Mathematics, Statistics and Computing
University of Greenwich, London SE18 6PF

First published 1989 by
MACMILLAN PRESS LTD
Houndmills, Basingstoke, Hampshire RG21 6XS
and London
Companies and representatives
throughout the world

ISBN 0-333-45935-0

A catalogue record for this book is available
from the British Library.

12 11 10 9 8 7 6
02 01 00 99 98 97

Printed in Hong Kong

CONTENTS

EDITOR'S FOREWORD

Wide concern has been expressed in tertiary education about the difficulties experienced by students during their first year of an undergraduate course containing a substantial component of mathematics. These difficulties have a number of underlying causes, including the change of emphasis from an algorithmic approach at school to a more rigorous and abstract approach in undergraduate studies, the greater expectation of independent study, and the increased pace at which material is presented. The books in this series are intended to be sensitive to these problems.

Each book is a carefully selected, short, introductory text on a key area of the first-year syllabus; the areas are complementary and largely self-contained. Throughout, the pace of development is gentle, sympathetic and carefully motivated. Clear and detailed explanations are provided, and important concepts and results are stressed.

As mathematics is a practical subject which is best learned by doing it, rather than by watching or reading about someone else doing it, a particular effort has been made to include a plentiful supply of worked examples, together with appropriate exercises, ranging in difficulty from the straightforward to the challenging.

When one goes fellwalking, the most breathtaking views require some expenditure of effort in order to gain access to them; nevertheless, the peak is more likely to be reached if a gentle and interesting route is chosen. The mathematical peaks attainable in these books are every bit as exhilarating, the paths are as gentle as we could find, and the interest and expectation are maintained throughout to prevent the spirits from flagging on the journey.

Lancaster, 1988

David A. Towers
Consultant Editor

PREFACE

Whilst there are a number of recent texts in mathematical modelling of the 'case study' variety, these are generally of most use as source material for the teacher. This *Guide to Mathematical Modelling*, however, is intended to be read by students; so the topics treated and the order of contents have been chosen with this in mind. We have tried to address the problem of how mathematical modelling is done as well as what a mathematical model is, and so have avoided presenting just a long catalogue of completed modelling examples.

The book is essentially a first course; so the amount of prerequisite mathematics and statistics is quite modest. It is chiefly aimed at the first-year level in an undergraduate degree course in mathematical sciences, but the treatment is such that the book could be used in the second year of a school sixth form. The contents have formed the basis of the first-year modelling course for students studying for B.Sc. in Mathematics, Statistics and Computing at the University of Greenwich (formerly Thames Polytechnic) and have proved a successful component in this course. We also gratefully acknowledge the influence of the pioneering work of many colleagues from the Open University and the former Polytechnics in the area of teaching mathematical modelling. The book stops short of investigating large-scale simulation models requiring software packages, but it lays valuable groundwork for subsequent study of such models.

At the outset, it is important to explain not only what modelling is, but also why it is worth doing. It is not merely a means of making the usual first-year curriculum in mathematics and statistics more lively and applicable. To accept that is to miss the point. The objective is to provide an approach to formulating and tackling problems in terms of mathematics and statistics. Eventually, when entering employment where real problems have to be dealt with, mathematicians will require additional skills to those fostered by study of conventional topics on the curriculum. The study of modelling promotes the development of these extra skills.

The book is divided into 10 chapters. Although it is not necessary to read the book strictly in chapter order, this may be preferred since there is some progression in difficulty as the subject is developed. It is vital, however, that readers try their hand at solving many of the problems posed, since modelling skills can only be learned by active participation.

Having set the scene in the opening chapter, some simple modelling problems are presented in chapter 2. These come from a variety of backgrounds, and readers should try some of the examples themselves from the problem descriptions provided. Mathematical modelling is by its nature difficult to structure, but it is useful to lay down general guidelines within which to operate when faced with new situations. To this end a general methodology is described in chapter 3.

The succeeding three chapters are particularly important for the beginner. Here the essential skills for successful modelling are developed. These are as follows.

1 Identifying the problem variables.
2 Constructing appropriate relations between these variables.
3 Taking measurements and judging the size of quantities.
4 Collecting data and deciding how to use them.
5 Estimating the values of parameters within the model that cannot be measured or calculated from data.

The backbone of the text comes in chapters 7 and 8. Chapter 7 deals with approaches to problems involving random features which demand some statistical analysis. Chapter 8 covers modelling situations which give rise to differential equations, such as are often encountered in physics and engineering.

Communication is vital for successful implementation of a mathematical model. It is necessary to explain ideas behind a model to other people, some of whom may not necessarily hold the same opinion as the modeller. It is also necessary to advise on the use of a model, often to non-specialists who need only to understand the essential points. Further, both at college and later in employment, it is often necessary to present findings verbally to a small group. These communication skills do not always come naturally; so, in chapter 9, advice is given on these matters.

Finally, in chapter 10, more demanding modelling assignments are presented. Some of the models are fully developed but others are left for the reader to process.

The content of this book complements other material usually studied in a mathematics degree course, and there is plenty of scope for further work in modelling as experience in mathematics and statistics is increased. Solving real problems by mathematical modelling is a challenging task, but it is also highly rewarding. If by working through the book readers gain confidence to take up this challenge, then the authors will be satisfied that the effort of writing the book has been worthwhile.

Woolwich, 1988

D.E.
M.J.H.

1 WHAT IS MODELLING?

1.1 INTRODUCTION

In industry and commerce the availability of fast and powerful computers has made it possible to 'mathematise' and 'computerise' a range of problems and activities previously unsolvable owing to their complexity. The opportunities for application of mathematics and statistics have therefore increased over the last 25 years. This means that there are more careers in industry and commerce which require a mathematical and statistical input. When the impact of complementary and related skills in computing and information technology is taken into account, there is now a very large and varied field of employment.

These opportunities can only be met if there are enough newly qualified professionals available with the right qualities to contribute. Precisely what qualities you need to develop in order to become an effective user and applier of mathematics will be described in the succeeding chapters of this book.

It is important to realise at the outset that learning to apply mathematics is a very different activity from learning mathematics. The skills needed to be successful in applying mathematics are quite different from those needed to understand concepts, to prove theorems or to solve equations. For this reason, this book is bound to appear different from a text dealing with a particular branch of mathematics. There is no theory to learn and there are only a few guiding principles. This is not to suggest, however, that mathematical modelling is an easy subject. The difficulty is not in learning and understanding the mathematics involved but in seeing where and how to apply it. There are many examples of very simple mathematics giving useful solutions to very difficult problems, although generally speaking the complexity of the problem and of the required mathematical treatment go hand in hand.

Professional modellers have to deal with a variety of real problems, and their main task is to translate each problem into a mathematical form. This is the essence of modelling, and it can involve discussions to clarify the problem, identification of problem variables, estimation, approximation and advocation of courses of action that may cost money and time.

The power industry provides many examples of how mathematical modelling is used. Problems of flow of water, electricity, gas and oil, and the necessity to match provision of these with varying demand, clearly lend themselves to mathematical treatment. The risk aspects of power provision, much publicised these days in connection with nuclear power, are also analysed by use of statistical and mathematical models.

Another activity in which mathematical modelling plays an important role is planning. Many national and local government departments depend on mathematicians to predict, for example, changes in transport, education and leisure requirements as populations shift or change in structure.

1.2 MODELS AND MODELLING

Consider the problem of optimising traffic flow near a roundabout. Unless mathematical and statistical techniques are used at the planning stage to predict the flow of traffic, the alternative is to build several differently designed roundabouts at considerable expense in order to find out which is the best. There are many situations like this where the use of mathematics provides valuable information concerning the behaviour of a system at much lower cost than the alternative 'trial-and-error' approach.

In road and traffic research laboratories, many traffic flow situations are analysed theoretically. Data are collected on the speed, size and manoeuvrability of vehicles, traffic density and junction configurations. Relations between the essential variables are then drawn up using mathematical and statistical techniques. From examination and interpretation of the results the best roundabout configuration at some particular junction can be predicted. We say that the researchers have 'built a model' of the roundabout —not a physical model but a mathematical model. The model would usually be converted into a computer simulation, which could then be used to evaluate other similar roundabout designs. Other people within the laboratory will be capable of using the model, but it is the skill of constructing the model in the first place which we wish to capture.

Enough has been said now to give you an idea of what mathematical modelling is and why it is so important. Exact definitions are not essential at this stage, but you will notice that it is easier and more useful to explain the process of modelling than to ponder on exactly what we mean by a model.

Any model (including a physical model) can be defined as a simplified representation of certain aspects of a real system. A *mathematical model* is a model created using mathematical concepts such as functions and equations. When we create mathematical models, we move from the real world into the abstract world of mathematical concepts, which is where the model is built. We then manipulate the model using mathematical techniques or computer-aided numerical computation. Finally we re-enter the real world, taking with us the solution to the mathematical problem, which is then translated into a useful solution to the real problem. Note that the *start* and *end* are in the real world.

It is also important to realise at the outset that mathematical modelling is carried out in order to solve *problems*. The idea is not to produce a model which mimics a real system just for the sake of it. Any model must have a definite purpose which is clearly stated at the start. This statement may itself vary according to the point of view of the model user and there could in some cases be a clash between opposing groups regarding the particular objectives involved. For example, the effect of a new road bypass on a town centre traffic jam could be viewed differently depending on whether we side with the drivers, the pedestrians, the shopkeepers or the persons over whose land the new road will be built. It is possible that the model construction will be affected by such viewpoints, although generally it will not be the modeller's job to make a moral judgement on the issues.

It must not therefore be thought that for a particular problem there is one right and proper model. We are not in the same situation as with arithmetic or algebra, where, to each question, there is one correct answer. Many different models can be developed for tackling the same problem. (It is also true, and a remarkable demonstration of the power of mathematics, that the same abstract model can often be used for quite different physical situations.) Some models may be 'better' than others in the sense that they are more useful or more accurate, but this is not always the case. Generally the success of a model depends on how easily it can be used and how accurate are its predictions. Note also that any model will have a *limited range of validity* and should not be applied outside this range.

1.3 THE LEARNING PROCESS FOR MATHEMATICAL MODELLING

It is easy to describe real modelling problems undertaken by the professionals, but how are you to begin to develop your own expertise? As you start to read this book, you are probably reasonably confident about elementary calculus, algebra and trigonometry and perhaps also some statistics and mechanics, but constructing mathematical models is a different matter. This book aims to help you to learn how it is done. It is not necessary to attempt complicated modelling problems based on industrial applications. The 'art and craft' of model building can be learned by starting with quite commonplace situations which contain a mathematical input based only on the mathematical work done at secondary-school level. As experience and knowledge are gained both in conventional mathematics and statistics as well as in modelling, then increasingly demanding problems can be considered. The first examples need not be contrived or false, for there are plenty of simple real-life situations available for study. Within chapter 2, 10 such examples are investigated.

By the time that you have worked your way through to the end of this book, you should have gained considerable experience of mathematical models and modelling. It is important to *do* modelling yourself, to try out your own ideas and not to be afraid to risk making mistakes. Learning modelling is rather like learning to swim or to drive a car; it is no good merely reading a book on how to do it. Similarly, with modelling, it is not sufficient to read someone else's completed model. Also mathematics has perhaps acquired a reputation for being a very precise and exact subject where there is no room for debate: you are either right or wrong. Of course, it is entirely appropriate and necessary that mathematical principles are based on sound reasoning and development but, when we come to *model* some given problem, we must feel free to construct the model using whatever mathematical relationships and techniques seem appropriate, and we may well change our minds several times before we are satisfied with a particular model.

It is often important, for the best results, *not* to work on your own. In industry, it is normal for a team of people to work together on the same model, and the team may consist of engineers or economists as well as mathematicians. It should be the same for beginners at the student level; although you may read this book by yourself, we hope that most of the modelling exercises are tried amongst a group. Different people have different suggestions to make, and it is important to pool ideas.

To be a succesful mathematical modeller it is not sufficient to have expertise in the techniques of mathematics, statistics and computing. Additional skills have to be acquired, together with the following general qualities: clear thinking, a logical approach, a good feel for data, an ability to communicate and enthusiasm!

1.4 SUMMARY

1 Mathematical modelling consists of applying your mathematical skills to obtain useful answers to real problems.
2 Learning to apply mathematical skills is very different from learning mathematics itself.
3 Models are used in a very wide range of applications, some of which do not appear initially to be mathematical in nature.
4 Models often allow quick and cheap evaluation of alternatives, leading to optimal solutions which are not otherwise obvious.
5 There are no precise rules in mathematical modelling and no 'correct' answers.
6 Modelling can be learnt only by *doing*.

2 GETTING STARTED

2.1 INTRODUCTION

As we have said in chapter 1, modelling is an active pursuit which you learn best by doing yourself or in a small group. A variety of examples have been selected for 'getting started' and, while each has been developed in the text, it may be preferable to try some first directly from the problem descriptions given. On the other hand, you may be more at home after seeing how they are tackled, and in any case you may wish to check the results and 'solutions'. We must be careful here about using terms such as 'solution', since in modelling there are many cases where there is no single solution in the conventional manner. Also the term 'problem' will be used as well as 'model' and you may wonder whether there is any difference. We shall return to this point in chapter 3; in the meantime, we shall set these semantics to one side.

2.2 EXAMPLES

The 10 examples given are intended to show what modelling is about in practice. They are not all equally difficult or necessarily of equal length. The amount of mathematical technique required to support the examples is quite modest. Within some, a little support is required from more advanced topics in mathematics and statistics. (Those topics which are probably being studied alongside a first course in modelling.) Alternative approaches can be found to those given in most of the examples, and this is quite usual in mathematical modelling. As we shall see, it is also quite normal to have 'second thoughts' and to want to improve a 'solution', perhaps using more sophisticated techniques than those used here. This progression is to be encouraged, and the way that it can be done is described in the next chapter. Thus, although some of the examples can be easily extended to investigate the situations in greater depth, each has been treated initially in a straightforward manner.

Example 2.2.1: Tape

Problem description

A common problem in lagging water pipes or bandaging an injured limb is to produce a neat job without too much overlap of material. The amount of lagging required to cover a certain length of pipe will be important, but also of interest is whether there is some relation between the bandage (or tape) width, the diameter of the pipe and the angle of pitch of the tape when it is wound around so as to make a neat join. The relation between these quantities is not obvious and this is what we shall set out to find.

There are two approaches to situations such as this.

1 To try a theoretical method to deduce a formula.
2 To collect data by carrying out measurements and to attempt to obtain a formula by drawing graphs.

To proceed with either approach, the problem needs to be clearly stated and the problem variables identified. First, two assumptions will be made.

(a) All the pipes have a circular cross-section.
(b) The tape is wound so that no overlap occurs.

Now denote the tape width by W (cm), the angle of pitch by A (deg) and the circumference of a pipe by C (cm). This introduces the term 'angle of pitch', which is best explained by referring to Fig. 2.1.

Procedure

Leaving theory aside for the moment to concentrate on the practical approach, you need a sample of pipes from which measurements can be taken and a selection of tapes of different widths. A protractor is also needed to measure angles of pitch. Having made some measurements, the resulting data are then organised so that graphs can be drawn. Typical data are shown in Table 2.1. You can work with either the diameter or the circumference of a pipe.

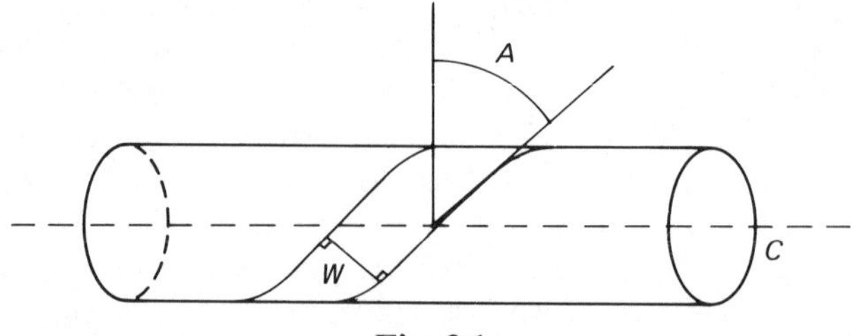

Fig. 2.1

Table 2.1

Circumference C/cm	10	10	10	10	10
Tape width W/cm	3	5	7	9	11
Angle of pitch A/deg	17	30	44	64	—
Circumference C/cm	20	20	20	20	20
Tape width W/cm	3	5	7	9	11
Angle of pitch A/deg	9	14	20	27	33
Circumference C/cm	30	30	30	30	30
Tape width W/cm	3	5	7	9	11
Angle of pitch A/deg	6	10	14	17	21

It may be difficult to deduce the required relation from graphs, especially if the measurements are not very accurate. If you are really baffled, refer to the theoretical method given at the end of the chapter, and then see whether your data fit in with the theory. Before doing this, however, think about what will happen when the tape width is either very small or very large. Carry out the resulting measurements. In particular, deduce what happens to A when

(a) W approaches zero and
(b) W becomes equal to the circumference C.

Now, as suggested just before the commencement of the examples, you may have some further ideas on the problem that you want to try out. A number of possible investigations will be listed at the end of all the examples in this chapter.

Follow-up

(a) Does it matter whether the pipe cross-sections are square or some other shape?
(b) Given a certain pipe length to lag with a particular tape supplied, how much tape do you need and what about the end effects?
(c) If the tape is put on with overlap, what effect does this have on the results?

Example 2.2.2: Fixtures

Problem description

The tennis club captain has to arrange the order of play for the annual club mixed-doubles tournament. There is only one court available and the tournament rules are that every pair has to play a match against every other pair. On a particular occasion, five pairs enter and the captain sets to work to decide the match order ready for a prompt 2.00 pm start. The captain's problems are as follows.

(a) How many matches will there be? The time scale may matter since the whole tournament must be over by the end of the day, and the last match must be played before it gets dark.
(b) How shall the order of play be arranged so that each pair has a reasonable rest between games?

Procedure

Here is an 'organisational' problem in which, unlike Example 2.2.1 above, no measurements are needed. The first part of the problem is straightforward apart from the decision on match length. If each match consists of three sets of tennis, the captain has the worry that matches could last a long time and the tournament will not finish in time.

The total number of matches to be played will be the number of selections of two pairs from the original five pairs. In mathematical terms, this is

$$\binom{5}{2} = \frac{5!}{2!3!} = 10.$$

Now a three-set tennis match can last for more than 1 h; so the captain's first problem is answered—he will have to shorten each match somehow since he does not want a ten-hour tournament.

The second problem is more difficult. The captain attempts to fix the order of play by trial and error, hoping that each pair will get an even distribution of matches throughout the day. The order can be displayed as in Table 2.2, where the pairs are represented by letters A, B, C, D and E.

Table 2.2

	A	B	C	D	E
A		1	6	9	3
B	1		4	7	10
C	6	4		2	8
D	9	7	2		5
E	3	10	8	5	

Table 2.3

Pair	Length of break (in matches)		
A	1	2	2
B	2	2	2
C	1	1	1
D	2	1	1
E	1	2	1

The interpretation is that pair A play in matches 1, 3, 6 and 9 and so on for B, C, D and E. This guesswork procedure seems to provide a reasonable distribution with all pairs getting a break between matches. However, the breaks are not evenly spread between the pairs, as can be seen in Table 2.3.

This plan, however, could be upset by a last-minute change of circumstances —suppose that one of the pairs withdraws or an extra pair makes a late entry. The club captain may wish that there was a more systematic way of fixing the order of play.

Follow-up

(a) For the five pairs above, is there an alternative order which is more satisfactory in that the breaks between matches are evenly spread? (This is quite difficult to achieve in general.)

(b) The club captain is now appointed fixtures secretary for the local tennis league, which operates with eight clubs per division. Each club is to play one match per week against opponents from its own division. Can you help by constructing the fixtures?

(c) The following year's club tournament is more popular with eight pairs entering. A whole day is set aside and there are now two courts available. Can you produce an order of play?

(d) In a particular inter-club match where there are three pairs in each team, it may happen that each team fields four men and two ladies. How would the matches be arranged so that everyone plays in three matches which are men's, ladies' or mixed doubles?

Example 2.2.3: Ferry

Problem description

A common problem for river ferry operators is the positioning of vehicles on the ferry deck so that the maximum number of vehicles can be safely parked. Unlike the situation for cross-channel ferries, data about the numbers of cars and lorries will not be known in advance. This means that the river ferry operators do not generally bother to sort out separate areas for cars and lorries but just load up on a 'first come, first served' basis. Using the following data, we want to see how this situation works in practice.

(a) The vehicle deck is 32 m long and can take two columns of vehicles side by side.
(b) Vehicles arrive at random and form a single queue whilst waiting for loading instructions.
(c) On average, 40% of the vehicles are cars, 55% are lorries and 5% are motor cycles.
(d) The car length varies between 3.5 m and 5.5 m, while the lorry length is between 8.0 m and 10.0 m.

The problem is to work out how many vehicles will be carried, whether they are cars or lorries and how much wasted space there is. There is the element of *randomness* and *queueing* in the situation to be investigated, and this is common in many modelling problems. Random-number models are dealt with in fuller detail in chapter 7, but we shall try to 'solve' this ferry problem now in order to give an idea of what is involved.

Procedure

All that is needed is a stream of random numbers and plenty of paper. There are *decisions* to be made as follows.

1 Is the next vehicle to board a car or lorry?
2 What is the length of the vehicle?
3 Which of the two parking columns does the vehicle join?
4 Do motor cycles matter?

Random numbers are available from tables, uniformly distributed on the interval (0, 1) and can be used to deal with questions 1–4. All that is needed is to be able to interpret from a random number RND on (0, 1) to decide which vehicle is next to board and also what length it is. As a general rule, if we require a random number on (a, b) rather than on (0, 1), then the simple transformation

$$a+(b-a)\text{RND} \tag{2.1}$$

will achieve this, where RND is read from tables. You will notice that this means we are assuming that the lengths of the cars and lorries are uniformly distributed over the given ranges. If the car lengths are more likely to be close to some average, then a more sophisticated statistical distribution will be needed (see chapter 7).

Thus, two random-number streams are required – one to decide what sort of vehicle is being loaded, and then a second stream to decide the lengths of

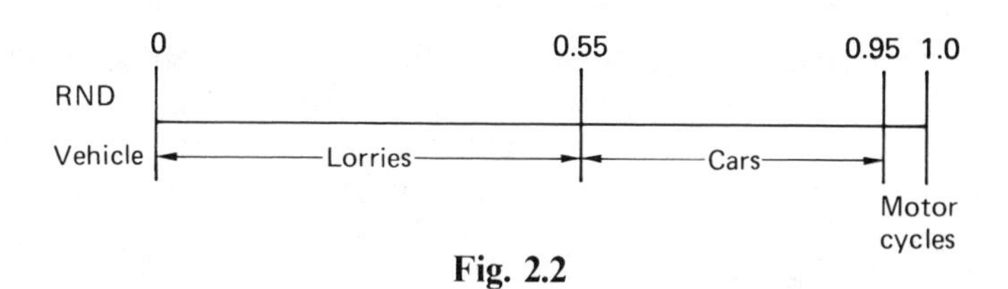

Fig. 2.2

each vehicle. To decide whether the current vehicle about to enter the ferry is a car or lorry, it is convenient to use the scale line drawn in Fig. 2.2. This shows the given percentages of cars, lorries and motor cycles marked off.

The lengths of the vehicles are determined from equation (2.1):

$$\text{car length} = 3.5 + 2.0 \text{ RND},$$

$$\text{lorry length} = 8.0 + 2.0 \text{ RND}.$$

Thus, supplying the two streams of random numbers will allow the loading data to be calculated, and a typical column length on the ferry will have been represented. For initial simplicity, only one line of parked vehicles will be considered. The situation as it arises is recorded in Table 2.4.

What happened to the motor-cyclists? They do not seem to have been considered. However, on short river crossings, we shall assume that any motor cycle can be fitted in alongside the vehicles and so does not take up any space at all in the column.

Now there is a snag with the problem investigation as it has so far been developed. The problem posed at the start was to examine how many cars were carried and whether there was much wasted space. The data in Table 2.4 give us a result but it is based on the inherent randomness described in the vehicle arrivals and lengths. In other words, for different streams of random numbers, there will be different results! What is actually required from this investigation (for possible use by the ferry operators) is the answer to such questions as 'on average how much wasted space is there when vehicles

Table 2.4

RND	0.10	0.28	0.61	0.34	0.77	0.57	0.02	0.88
Vehicle	Lorry	Lorry	Car	Lorry	Car	Car	Lorry	Car
RND	0.59	0.48	0.10	0.56	0.30	0.90	0.81	0.66
Vehicle length/m	9.44	8.96	3.70	9.12	4.10	5.30	9.62	4.82
Column length/m	9.44	18.4	22.1	31.22	35.32	40.62	50.24	55.06

are allowed to drive onto the ferry in the first come, first served manner?' We *can* answer this question with a little work, but the key words are 'on average'. To obtain this average value, we need to repeat the above calculations several times, which is why a lot of paper will be needed! On the other hand, the process really needs to be made automatic through the use of a computer program.

At this early stage the best way forward is to share the investigation around between a number of students all operating independently, so that you only do one set of calculations yourself. A value for the required average can then be calculated.

Follow-up

(a) Now try two columns of vehicles as originally intended, either segregated or mixed.
(b) Compare the waste of space incurred between the two strategies of segregated loading and non-segregated loading.
(c) Consider a cost factor based on the number of paying passengers–there are probably several people in each car, but only the driver in each lorry.

Example 2.2.4: Home decorating

Problem description

A popular do-it-yourself task is wallpapering. Assuming that the type of wallpaper has been chosen, then you have to know how many rolls of paper to buy. The measurements of the dimensions of the room can be taken, but you now need to know how to convert these data into the number of rolls. An overestimation can be made but the paper may be expensive and money

Table 2.5

Height of room/ft	Number of rolls required for the following distances around walls including doors and windows						
	32 ft	36 ft	40 ft	44 ft	48 ft	52 ft	56 ft
$7-7\frac{1}{2}$	4	5	5	6	6	7	7
$7\frac{1}{2}-8$	4	5	5	6	6	7	8
$8-8\frac{1}{2}$	5	5	6	6	7	7	8
$8\frac{1}{2}-9$	5	5	6	6	7	8	8
$9-9\frac{1}{2}$	5	6	6	7	7	8	9
$9\frac{1}{2}-10$	5	6	7	7	8	9	9
$10-10\frac{1}{2}$	5	6	7	8	8	9	10

is then wasted. To enable this calculation to be carried out, most do-it-yourself shops have a leaflet containing a table of numbers arranged in a block rather like a matrix.

Table 2.5 is intended to show how many rolls are needed, given the room size.

The number in the main body of the table indicates how many rolls are needed. It is interesting that such a simple chart can provide an accurate result; so you may feel that a check of the table entries is necessary. Additional information is available that all wallpaper rolls are 10 m long and 52 cm wide; also the pattern repeat is 35 cm.

How can you check the accuracy of the table? Suppose that the dimensions of the room to be decorated are 12 ft wide, 14 ft long and 9 ft high. There is one door which is 85 in long by 42 in wide including the frame. There is only one window 50 in wide and 5 ft high from the ceiling downwards.

Procedure

Carry out some easy calculations as follows:

$$\begin{aligned}\text{total wall area} &= (12 \times 9 \times 2 + 14 \times 9 \times 2)\ \text{ft}^2 \\ &= 468\ \text{ft}^2\end{aligned}$$

and

$$\begin{aligned}\text{area of door and window} &\approx (7 \times 3.5 + 5 \times 5)\ \text{ft}^2 \\ &\approx 50\ \text{ft}^2.\end{aligned}$$

Hence

$$\text{wall area to be papered} = 418\ \text{ft}^2. \tag{2.2}$$

Now one roll of wallpaper has

$$\begin{aligned}\text{length} &= \frac{10 \times 39.37}{12}\ \text{ft} \\ &\approx 32.8\ \text{ft}\end{aligned}$$

and

$$\begin{aligned}\text{width} &= \frac{52}{2.54 \times 12}\ \text{ft} \\ &\approx 1.71\ \text{ft}.\end{aligned}$$

(The unit conversions are given in chapter 4.) So one roll of wallpaper covers an area given by

$$\begin{aligned}\text{area} &= 32.8 \times 1.71\ \text{ft}^2 \\ &\approx 56.09\ \text{ft}^2. \end{aligned} \tag{2.3}$$

Finally, by division,

$$\text{number of rolls required} = \frac{418}{55.96}$$

$$= 7.45.$$

Now return to Table 2.5 and read off the value for

$$\text{number of feet round wall} = 52,$$

$$\text{room height} = 9.$$

So, according to Table 2.5, the number of rolls needed is 8, which is in good agreement with the actual calculation.

Follow-up

(a) You decide to paper the ceiling. How many rolls of ceiling paper are needed for the room, assuming the rolls of ceiling paper are the same size as wallpaper? Draw up a table for general use in selling ceiling paper.
(b) Home decorating also requires some painting. A gloss finish is the most important feature and so how much paint should be bought? (1 litre of gloss paint covers 17 m^2 with one coat.)

Example 2.2.5: Traffic lights

Problem description

How many cars can pass through a set of traffic lights when they turn green for a period of 15 s? Suppose that a line of cars are queueing to pass through a junction and the lights are set at red. Often the queue is so long that a driver some distance from the front of the queue sees the lights go green and back to red again before even moving forward himself, let alone actually getting across the junction. The way in which traffic behaves at lights can be quite complicated, depending on such things as 'right-turn filters', 'no left turn' and so on. Also, when the traffic density decreases, often no queueing takes place at all, and the number of vehicles passing across a junction will depend on how many there are in the vicinity of the traffic lights in the first place. Here we investigate the saturation case where there is already a long queue of cars poised at a red light waiting to move off.

A number of assumptions will be made at the outset.

(a) The junction is not blocked in any way.
(b) All cars intend to pass directly across the junction.
(c) All vehicles are cars of the same size, 5 m in length, and initially at rest.
(d) There is a 2 m gap between each stationary car.

The objective is to calculate how many cars pass across the junction during one green light cycle.

Procedure

How does a car pull away? We assume that it will accelerate uniformly up to a maximum speed of 30 miles h^{-1} (the town speed understood by most drivers if not actually observed!) We have a mixture of units here; miles per hour must be converted to metres per second:

$$\begin{aligned}30 \text{ miles h}^{-1} &= \frac{30}{3600} \text{ miles s}^{-1}\\ &= \frac{30 \times 5280}{3600} \text{ ft s}^{-1}\\ &= \frac{30 \times 5280 \times 12}{3600 \times 39.37} \text{ m s}^{-1}\\ &\approx 13.4 \text{ m s}^{-1}.\end{aligned}$$

Reasonably, the final speed will be taken as 15 m s^{-1}. Next, the acceleration of a car in city conditions has to be assessed. Some car manuals claim '0 to 60 miles h^{-1} in 10 s'. Using this, we get 60 miles h^{-1} = 26.8 m s^{-1} achieved in 10 s; so the acceleration is 26.8/10 m s^{-2} = 2.68 m s^{-2}. Acting a little conservatively, we shall take the acceleration to be 2.0 m s^{-2}.

On the assumption that, once the final speed has been reached, this speed is then held constant, we can draw a velocity–time graph (Fig. 2.3).

Hence after 7.5 s the car has reached its final speed of 15 m s^{-1}. Now describing the motion of one car is not sufficient for the objective of the model—the motion of the rest of the queue must be considered as well. Suppose there is a delay of 1 s in thinking time before the next car

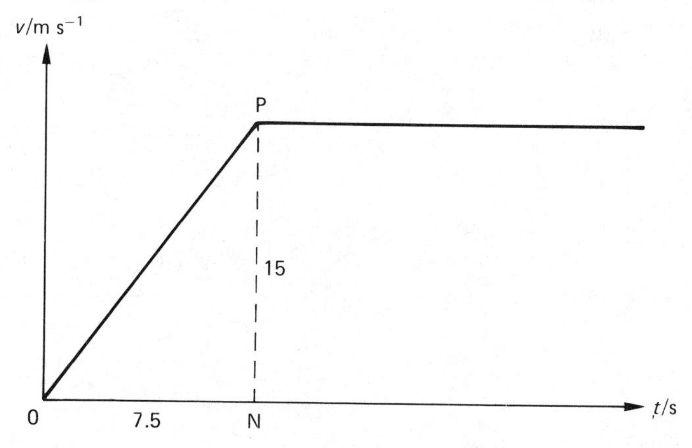

Fig. 2.3

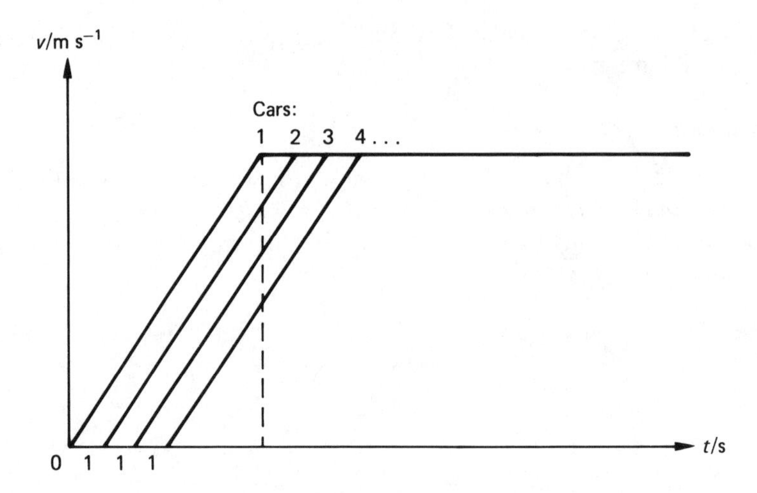

Fig. 2.4

reacts and pulls away. This motion is easily incorporated into the same velocity–time graph, and all following cars can be similarly marked in (Fig. 2.4).

The distance travelled by a car can also be conveniently evaluated from a velocity–time graph by calculating the area under the graph. Thus, after 4 s, we have

car 1 has gone	16 m,
car 2 has gone	9 m,
car 3 has gone	4 m,
car 4 has gone	1 m,
car 5 is just about to move off.	

However, where are these cars in relation to road markings at the traffic lights? Measuring from the front bumper of each car, we can draw up a table of car positions, both initially and after 4 s (Table 2.6).

Table 2.6

Car	Time in motion/s	Distance travelled/m	Initial position related to 'white line'/m	Net position related to 'white line'/m
1	4.0	16.0	0.0	16.0
2	3.0	9.0	−7.0	2.0
3	2.0	4.0	−14.0	−10.0
4	1.0	1.0	−21.0	−20.0
5	0.0	0.0	−28.0	−28.0

Now after 7.5 s the situation changes since some cars will have reached their maximum speed of 15 m s^{-1}. Can you work out a general formula for the distance s metres that each car has travelled after, say, T seconds?

A suggested formula is

$$s = \begin{cases} T^2, & T \leqslant 7.5, \\ 7.5^2 + 15(T - 7.5), & T \geqslant 7.5. \end{cases}$$

Next suppose that the lights stay *green* for 15 s; can you calculate how many cars pass through the junction? The situation at $T = 15$ is shown in Table 2.7.

Table 2.7

Car	Time in motion/s	Distance travelled/m	Initial position in relation to 'white line'/m	Net position in relation to 'white line'/m
1	15	168.75	0.0	168.75
2	14	153.75	−7.0	146.75
3	13	138.75	−14.0	124.75
4	12	123.75	−21.0	102.75
5	11	108.75	−28.0	80.75
6	10	93.75	−35.0	58.75
7	9	78.75	−42.0	36.75
8	8	63.75	−49.0	14.75
9	7	49.00	−56.0	−7.00
10	6	36.00	−63.0	−27.00
11	5	25.00	−70.0	−45.00
12	4	16.00	−77.0	−61.00
13	3	9.00	−84.0	−75.00
14	2	4.00	−91.0	−87.00
15	1	1.00	−98.0	−97.00
16	0	0.00	−105.0	−105.00

You can use the table to calculate how many cars pass through before the lights are set at *red* again. (The answer is given at the end of the chapter.)

Follow-up

(a) Can you model how the cars will draw to rest again as the lights turn red? This depends on 'braking times', and data on this can be taken from the *Highway Code* booklet (Table 2.8).

What is the relation between *speed* and *total distance* expressed as a mathematical formula? Can this be incorporated into the model?

(b) In a very long queue waiting at traffic lights, there will be considerable stopping and starting as the effect of moving up the queue is passed along. Is it possible to model this 'wave' effect as time goes by?

Table 2.8

Speed/miles h^{-1}	Thinking distance/ft	Braking distance/ft	Total distance/ft
20	20	20	40
30	30	45	75
40	40	80	120
50	50	125	175

Example 2.2.6: Price war

A common occurrence in retailing is competition between rival establishments over which shop can claim most of the market. Market share can be affected by advertising locally, by price variation and by sales gimmicks, etc. The most straightforward price war situations occur in supermarkets or in garages. Here a petrol sales price war is considered.

Problem description

Two petrol stations operate from adjacent main road sites and vie with one another for business. Competition is also stiff from other petrol stations not far away and profit margins from the sale of petrol are very sensitive to sudden changes in demand. On the other hand, the market is large and, although both petrol stations have regular customers, they realise that many of their sales are to 'casual' drivers.

One day, one of the petrol stations suddenly drops the 'price per litre' as advertised on the station forecourt in an attempt to attract more of the market. The other petrol station immediately notices a fall in trade as a result. They decide that they must follow suit in dropping the price and so start up a '*price war*' with the first garage. The second garage wants to devise a strategy so that, in altering their price, earnings will be as high as possible.

Procedure

Let the first garage be denoted by A, and the second by B. The crucial part of setting up a model here is to attempt to predict how B's market share will be affected by the sudden price drop by A. We have also to take account of the 'normal price' of petrol still being charged by other garages. To quantify these ideas, we introduce some notation:

let the normal selling price be p pence/litre,

let A's new selling price be y pence/litre,

let B's new selling price be x pence/litre,

let B's steady sales volume be (before the price war starts) L litres/day,

let the cost price of petrol be w pence/litre.

Now this example is typical of a more 'open' situation where we have to speculate on the behaviour and relation between variables. This is one of the skills that we are going to need if we are to become effective mathematical modellers.

To build a model of sales volume variation, we now assume that B's daily sales are affected as follows. Sales are changed by an amount directly proportional to the following.

1 The difference between B's price and A's price.
2 The difference between B's price and the normal price.
3 The difference between A's price and the normal price.

Thus our model for the new daily sales for B is

$$L - a(x-y) - b(p-y) + c(p-x),$$

where a, b and c are proportionality constants. Note that a, b and c must all be positive. An 'earnings function' can now be calculated for garage B, since earnings are given by 'price × sales volume minus costs'. Hence the daily earnings of B are

$$E(x, y) = (x-w)[L - a(x-y) - b(p-y) + c(p-x)] \text{ pence.}$$

This expression can be maximised as a function of x, regarding y as an input parameter. It is easily shown by elementary calculus that the x value for maximum E is

$$x = \frac{L + y(a+b) - p(b-c) + w(a+c)}{2(a+c)}.$$

The model is now tried with data as follows:

$L = 20\,000$,

$p = 40$,

$w = 30$,

$y =$ various values such as 37, 38 and 39.

(Prices can vary owing to tax and market features.)

The proportionality constants are our main worry and have to be estimated in some way or other. (This often happens in mathematical modelling.) This particular model would need real data to enable a, b and c to be fixed, but at this stage we are concerned with *principles* as much as accurate data collection. It is quite instructive to judge what size of parameter to select for a, b and c. By examining the relative sizes of the quantities within the sales function, we decide to choose $a = 4 \times 1000$ and $b = c = 1000$. You should be able to see that taking $a = 1.0$, or 0.2, say, will not be sensible. Also we might expect that the terms containing a, b and c are not equally important, i.e. it is not necessary to choose the proportionality constants all equal. Using the above values provides us with sensible results, and the results can be summarised as in Table 2.9.

Table 2.9

Selection of y	Optimum x	$E/£$
39	36.5	2112.5
38	36.0	1950.0
37	35.5	1787.5

Note that original daily earnings $= (40 - 30) \times 20\,000$ pence

$$= £2000$$

Follow-up

(a) Criticise the formula for B's new daily sales. Is it the most appropriate? What is the effect on the formula when

(1) $y = p$,
(2) $y = x$ and
(3) $x = p$?

Does the effect fit in with common sense?

(b) How sensitive are the results to the choices made for the proportionality constants? Why were they chosen to be O(1000)? Try alternatives.

Example 2.2.7: Evacuation

From time to time, any organisation occupying a large building will want to practise an evacuation procedure in case of an emergency. The evacuation may be required for a variety of reasons but in all cases will have to be organised to proceed in an orderly manner.

The people whose responsibility it is to check the safety of the arrangements for evacuation will want to make sure that in each room there are clear instructions on how to leave the building. They will also want to have some idea of how long it will take for the building to empty. This will then enable them to decide whether the number of exit routes are sufficient to cope when the building is full to capacity.

To avoid having too many practice evacuations, it is helpful if an estimate for the total exit time can be made by trying a theoretical approach instead. This is where a useful modelling exercise can be developed.

Problem description

The above general issues are rather difficult to deal with since we shall need to know how large the building is, how many exit stairs are available, how many people per room and so on. However, a good modelling tactic is to consider a simplified problem when the original seems too hard or too undefined. Often the outcome from a simpler situation is contributory to the main problem anyway.

Consequently, we shall take the building to be a school or college and confine attention to one floor where there are a number of similar classrooms in a row. Suppose that each is evacuated by walking along the corridor serving the rooms and then through the exit door at the end. The situation is shown in Fig. 2.5.

The objective of the simplified problem is to work out how long it takes for the four rooms shown to be completely emptied and the occupants all to pass through the exit door.

Procedure

It is instructive first to consider an orderly departure from a single room. A list of the relevant variables helps to clarify the situation:

denote the number of people in the room by (one teacher and n_1 students)	$n_1 + 1$,
let the spacing between each person be	d (m),
assume that everyone walks at a constant speed	v (m s^{-1}).

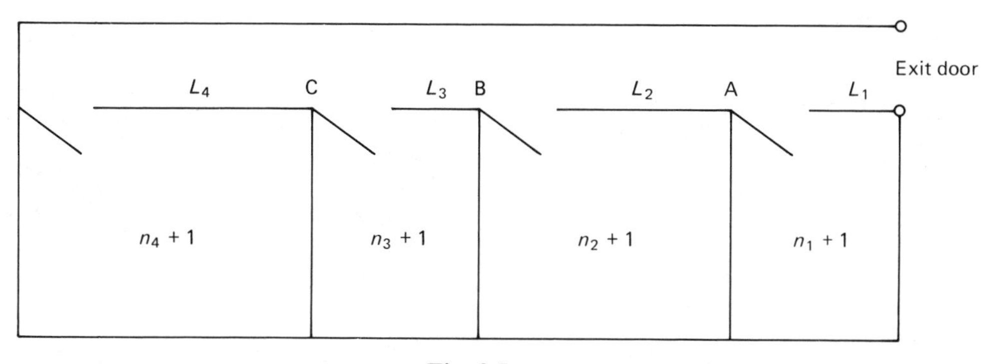

Fig. 2.5

Thus there is a chain of people of length $n_1 d$ m filing out of the room. On the assumption that the leader of the chain starts at the door, then the time needed for everyone to leave is $n_1 d/v$.

There will be an initial short delay while the first person reaches the classroom door. Denoting this by t_0, then we have the total room emptying time of $n_1 d/v + t_0$. At this point, the last person has just reached A in Fig. 2.5; so, as there is still a distance L_1 to walk to the exit door, an additional time of L_1/v is needed for this. Thus, finally the total time for the first room to evacuate is

$$\frac{L_1}{v} + \frac{n_1 d}{v} + t_0. \tag{2.4}$$

This is all quite straightforward, but we have only considered one room. Now refer to the second room, which could be a different size from the first and contain a different number of people. It is likely that t_0 will be the same, and v and d also the same. Given that there are $n_2 + 1$ people, then the chain of people for the second room takes $n_2 d/v + t_0$ for the last person to reach the door of the classroom.

The corridor length that this second chain has to walk along is $L_1 + L_2$ m, from point B to the exit. At a speed of v m s^{-1} the time needed for this is $(L_1 + L_2)/v$. The total evacuation time from the second room is

$$\frac{L_1 + L_2}{v} + \frac{n_2 d}{v} + t_0. \tag{2.5}$$

However, there is a fallacy in this analysis, since no account has been taken of the possibility that the two chains will conflict in the corridor. Normally, corridors are sufficiently wide for at least two columns of people to walk side by side. We shall suppose for safety, however, that in this situation there is only enough width for one column to occupy the corridor at a time. This means that a delay can occur at point A when the second chain is held up while the first is still leaving.

The time for the leader of the second chain to reach point A is $t_0 + L_2/v$ and the time for the last person in the first chain to reach point A is $t_0 + n_1 d/v$. Hence a delay occurs when

$$t_0 + \frac{n_1 d}{v} > t_0 + \frac{L_2}{v},$$

i.e.

$$n_1 > \frac{L_2}{d}. \tag{2.6}$$

On the assumption that there is a delay, then the second chain joins on the end of the first. The total time now is

$$t_0 + \frac{n_1 d}{v} + \frac{L_1}{v} + \frac{n_2 d}{v} = t_0 + \frac{(n_1 + n_2)d}{v} + \frac{L_1}{v}. \tag{2.7}$$

If no delay takes place, then n_1 is too small (for example) and the total time for evacuation is the result from equation (2.5), i.e.

$$t_0 + \frac{n_2 d}{v} + \frac{L_1 + L_2}{v}. \tag{2.8}$$

Finally, we shall calculate the evacuation time when the chains blend together without a delay or a gap, i.e.

$$n_1 = \frac{L_2}{d}. \tag{2.9}$$

Data will be taken as follows. Let $L_1 = 10$ m, $L_2 = 12$ m, $v = 2$ m s^{-1}, $t_0 = 3$ s and the space gap $d = 1$ m. Then the number of people in the first room is known from equation (2.9) as $L_2/d = 12$, which means that there are 13 people in the room. Suppose that the second room contains 31 persons so that $n_2 = 30$. The evacuation time can now be calculated from equation (2.8) as $3 + 30 \times 1/2 + (10 + 12)/2$ s $= 29$ s.

Follow-up

(a) Only two of the rooms have been considered. What are the logistics when all four of the original rooms shown in Fig. 2.5 contain people?

(b) Suppose that the corridor is wide enough for two columns of people to file along through the exit door in safety. Consider the new situation and calculate a time for evacuation given that $L_3 = 12$ m, $L_4 = 15$ m, $n_1 = 20$ and $n_2 = 30$.

(c) Do you think that the model is realistic? Suggest and carry out some improvements.

Example 2.2.8: Crossing the road

Problem description

The local highways and transport division of the council has been asked whether it can approve the setting up of an additional pelican crossing to help pedestrians to cross a road at a certain point. The road carries a moderate amount of traffic so that people have a habit of 'making a dash for it' instead of walking to the next authorised crossing, which is some distance away. The council cannot provide pelican crossings at every popular crossing point but decide that the particular point on the road should be considered for a pelican crossing if the average waiting time there is greater than 15 s.

The following questions seem relevant to the situation.

(a) How dense is the traffic?
(b) How quickly can a person cross?
(c) Are there two lanes of traffic without a central reservation?

(d) Does driver reaction time come into it?
(e) What about bad visibility or perhaps a bend in the road?
(f) Are the interests of drivers more important than those of pedestrians because the road is otherwise a fast and safe highway? (We must not forget that many pedestrians are also drivers.)

In considering these questions, there are some general issues which we have raised but cannot deal with in a simple first model. For example the question of whose interests take priority may be important here and may affect the way in which any data that are collected on traffic flow are used. However, we shall restrict the investigation to the straightforward analysis of crossing the road through a stream of traffic.

Procedure

To begin, suppose that there is a steady flow of vehicles proceeding along the road, one direction being considered only, for the moment. If the vehicles are all travelling at 40 miles h^{-1}, would you cross? You can estimate that it will take 5 s at least to make it. So do you cross or not? After a little thought, we realise that the data are not complete. It is density of traffic that is needed, since one solitary car going at 40 miles h^{-1} does not create a problem. On the other hand, density of flow is related to speed since two cars travelling at 40 miles h^{-1} would not 'close' to a few feet apart! However, this confuses the issue between *vehicle safety in convoy* and *whether it is safe to cross*. After all, densely packed traffic crawling at 5 miles h^{-1} could be easy to penetrate, just as a very thin flow of fast cars is also easy to penetrate.

A particular case will be considered where the flow rate is a steady 1000 cars h^{-1}. Would you be able to cross? Converting to seconds gives

$$\begin{aligned} 1000 \text{ cars h}^{-1} &= \frac{1000}{60 \times 60} \text{ cars s}^{-1} \\ &= \frac{1000}{3600} \text{ cars s}^{-1} \\ &= \frac{5}{18} \text{ cars s}^{-1}, \end{aligned} \tag{2.10}$$

i.e. five cars regularly pass by every 18 s; so there is one car every 3.6 s in a steady procession. Hence, if you need 5 s to get across, then you will not be able to make it.

To generalise the result, some symbols will be introduced. Let the steady flow rate be q cars s^{-1}. Therefore the time gap between cars is $1/q$ s. Suppose that the time needed to cross is T s; then the conclusion is that you cross if $T < 1/q$, or, reversing this, if $q < 1/T$.

Now this simple result is valid only if the traffic flow is at a constant steady rate. The question of how long you are kept waiting before crossing does not occur in the above model. Either you can cross or you cannot cross!

Consider now a more realistic model: traffic flow is random in the sense that, although the average flow rate is still q cars s^{-1}, it is not a regular steady flow, i.e. the time gap between cars is *distributed about* $1/q$ s. The simplest way to incorporate this random effect into the model is to use the 'exponential distribution' from statistics. This results in the fact that the sequence of time gaps between cars can be calculated from the formula $-(1/q)\ln(\mathrm{RND})$. It can be shown that the average value of this sequence will be $1/q$ as we require. (The use of the exponential distribution will be explained in more detail in chapter 7.)

Now, there will always be a minimum gap, say t_0, between cars due to car length. If the time gap between cars is denoted by T_{gap}, then we have

$$T_{\mathrm{gap}} = t_0 - \frac{1}{q}\ln(\mathrm{RND}).$$

Keeping the crossing time needed for pedestrians set at T means that you cross whenever

$$T < t_0 - \frac{1}{q}\ln(\mathrm{RND})$$

or

$$-\frac{1}{q}\ln(\mathrm{RND}) > T - t_0.$$

We shall see how this works in practice. The use of random numbers has already been explained in Example 2.2.3.

Let us consider a trial: take $t_0 = 2$ and the traffic flow to be 1000 cars h^{-1}. Hence $q = 5/18$ cars s^{-1} and so $1/q = 3.6$, as before.

Thus, we require the quantity

$$T_{\mathrm{gap}} = 2.0 - 3.6\ln(\mathrm{RND}). \tag{2.11}$$

Given a supply of random numbers, then T_{gap} can be calculated as follows.

RND	0.53	0.38	0.58	0.82	0.26	0.71	0.22
$T_{\mathrm{gap}}/\mathrm{s}$	4.3	5.5	4.0	2.7	6.8	3.2	7.5
RND	0.02	0.63	0.21	0.17	0.69	0.64	0.55
$T_{\mathrm{gap}}/\mathrm{s}$	16.1	3.7	7.6	8.4	3.3	3.6	4.2

If we take 5 s to get across the road, then this output can be used to see whether we are able to cross or not.

However, one important fact has been omitted, namely *when* exactly do you arrive at the kerb side? This will be at a random time as well and needs to be incorporated before you know whether there is a wait or not. Suppose that we examine a 60 s period and decide that you arrive at the kerb side at any moment within the 60 s. Then, by comparing the arrivals with the T_{gap} data, it will be possible to calculate how long a wait is necessary before crossing. It is only by amassing many data about kerb-side waiting times that an eventual decision on the pelican crossing can be made.

To simulate a random arrival time within a 60 s period, we simply use equation (2.1) in Example 2.2.3, where $a = 0$ and $b = 60$. The arrival time is then 60 RND, where $0 < \text{RND} < 1$. Thus over, say, five trials the arrival times are as follows.

Trial	A	B	C	D	E
RND	0.59	0.48	0.10	0.56	0.30
Arrival time/s	35.4	28.8	6.0	33.6	18.0

The T_{gap} data are now expressed cumulatively and for convenience listed on a time scale starting from 0 as shown:

T_{gap}(cum) 0, 4.3, 9.8, 13.8, 16.5, 23.3, 26.5, 34.0, 50.1, 53.8, 61.4, ...

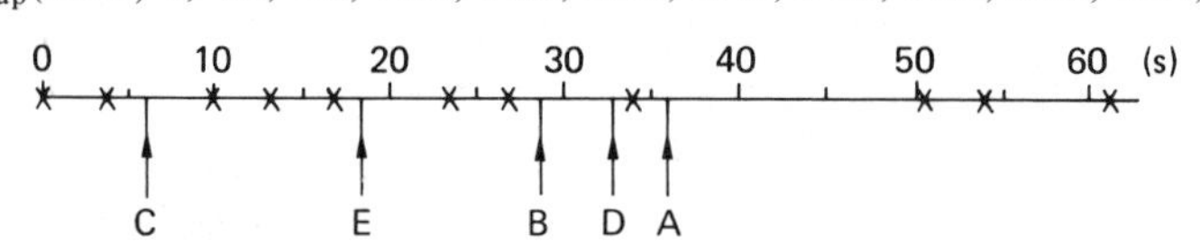

It is now easy to read off from the scale line the waiting times for each trial (Table 2.10).

In this way, data can be built up on how long we wait at the kerb. Clearly, many trials are needed before the average waiting time can be calculated. This model has the additional attraction that the theoretical data can be validated by actually standing at the point on the road under consideration and observing how long people wait.

Table 2.10

Trial	Arrival time/s	Time before next car/s	Decision
A	35.4	14.7	Go straight across
B	28.8	5.2	Go straight across
C	6.0	3.8	Wait 10.5 s
D	33.6	0.4	Wait 0.4 s
E	18.0	5.3	Go straight across

This kind of approach is typical with most models which involve random numbers. Such models are considered in more detail in chapter 7.

Follow-up

(a) Can you extend this to cope with two opposite directions of traffic flow?
(b) What happens if there is a central island?
(c) Given sufficient computing ability, you may like to write a program to calculate the waiting times automatically. This also helps when the input data on car flow density are altered.

Example 2.2.9: Corner

Problem description

A familiar scene in hospitals with a surgical unit is the moving of a patient from the ward to the operating theatre whilst the patient remains in bed. In other words the bed is pushed along the corridor by hospital staff at a fair speed without inconveniencing the patient. Unfortunately, some hospitals have narrow corridors with right-angle bends in them. Suppose that there is just such a bend between the operating theatre and the wards. Beds have to be wheeled along this corridor, negotiating the right-angle bend.

Often for maintenance and decorating purposes, ladders and long planks also have to be carried round the hospital corridors. Again there is a right-angle bend to negotiate.

There is a modelling problem in both these situations. It is interesting to see what length of ladder can be carried round the corner and also whether a typical hospital bed can be comfortably pushed round. In the interests of saving space, and for future corridor design, we might want to find out what the minimum width of a corridor is so that a bed can pass round a corner.

Procedure

A diagram (Fig. 2.6) is helpful in this situation.

It seems reasonable that the problem of moving a ladder round the corner is best considered first. Having dealt with a ladder, the bed can then be investigated. Another diagram (Fig. 2.7) is useful so that the ladder problem can be considered.

The problem is generalised so that, for corridor dimensions a and b and ladder of length l, what is the greatest value of l so that the ladder will still pass round the corner? (In practice the units of length will be taken in metres.) After some thought, we realise that what is required is the *minimum* value of PQ. From the trigonometry, $\text{PN} = \text{SN} \operatorname{cosec} \theta$ and $\text{QN} = \text{RN} \sec \theta$. Hence,

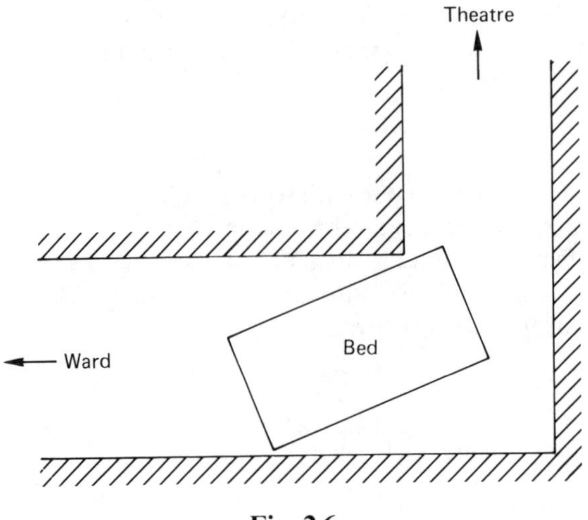

Fig. 2.6

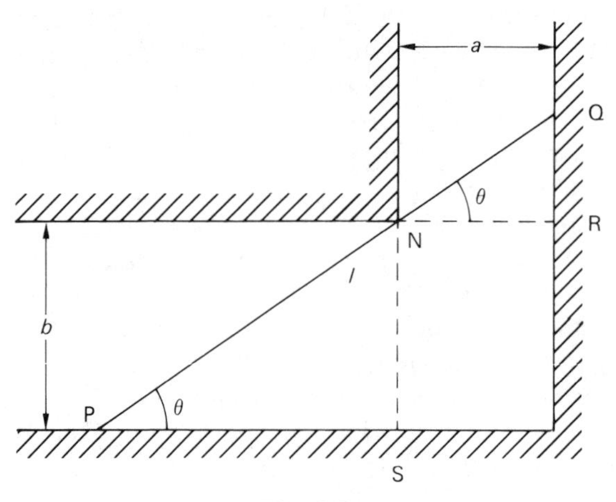

Fig. 2.7

by addition,

$$l = \text{PN} + \text{QN} = a \sec\theta + b \operatorname{cosec}\theta. \tag{2.12}$$

Using differential calculus,

$$\frac{\mathrm{d}l}{\mathrm{d}\theta} = a \sec\theta \tan\theta - b \operatorname{cosec}\theta \cot\theta.$$

On setting $\mathrm{d}l/\mathrm{d}\theta = 0$, the condition for a minimum is $\tan^3\theta = b/a$. This means that the ladder length we want can be calculated by substituting the θ value into equation (2.12). This gives

$$\text{length of ladder} = (a^{2/3} + b^{2/3})^{3/2}. \tag{2.13}$$

(Note that, to get the longest ladder for given a and b, it is the minimum value of l that is wanted.) Substituting values for a and b into equation (2.13) enables the ladder length to be calculated. Now let us return to the bed problem (Fig. 2.8).

First, the area A of the bed is given by $A = p \times q$. From the trigonometry in Fig. 2.8 a relation connecting p and q can be obtained using AB = AN + NB:

$$p = (a - q \sin \theta) \sec \theta + (b - q \cos \theta) \operatorname{cosec} \theta. \tag{2.14}$$

Hence,

$$A = q[(a - q \sin \theta) \sec \theta + (b - q \cos \theta) \operatorname{cosec} \theta]. \tag{2.15}$$

Although the objective here is to calculate minimum dimensions of a and b so that a standard bed can be pushed round the corner, it is more convenient as far as the mathematical model is concerned to tackle the problem the other way round and to treat p and q as variables for given a and b. In fact, p has been eliminated in equation (2.15); so we proceed with equation (2.15) regarding q and θ as variables. The approach is then analogous to that for the ladder, but the mathematics is a little harder. We attempt to minimise A in equation (2.15) using calculus but note that there are *two* independent variables to be differentiated, q and θ. Note that if $q = 0$, equation (2.14) becomes $p = a \sec \theta + b \operatorname{cosec} \theta$ (compare with (2.12)), and equation (2.15) becomes $A = 0$ as required.

It is easily shown in calculus textbooks on functions of several variables that the minimum is sought by differentiating with respect to each variable in turn, while keeping the other constant. This technique is known as 'partial differentiation' and will probably be covered in a parallel first-year course. To indicate a partial derivative, a slightly different notation ∂ for the derivative is normally used.

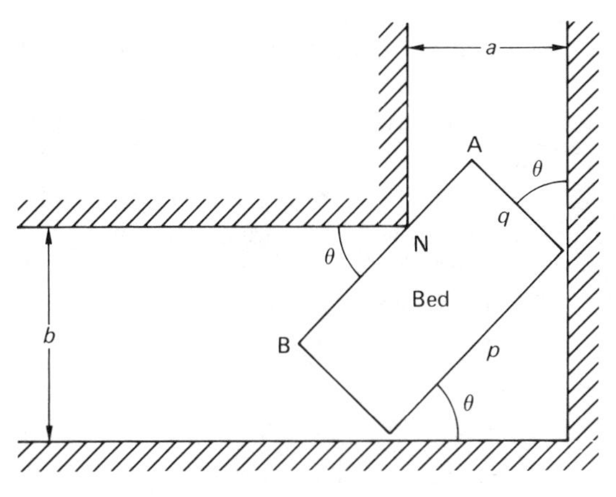

Fig. 2.8

Consequently, evaluating $\partial A/\partial q$ and $\partial A/\partial \theta$ gives

$$\frac{\partial A}{\partial q} = (a - q \sin \theta) \sec \theta + (b - q \cos \theta) \operatorname{cosec} \theta$$

$$+ (-\sin \theta \sec \theta - \cos \theta \operatorname{cosec} \theta) q,$$

$$\frac{\partial A}{\partial \theta} = [(a - q \sin \theta) \sec \theta \tan \theta - q \sec \theta \cos \theta$$

$$- (b - q \cos \theta) \operatorname{cosec} \theta \cot \theta + q \sin \theta \operatorname{cosec} \theta] q.$$

The required conditions for a minimum value of A are found by equating both these derivatives to zero. After a little manipulation the conditions reduce to

$$a \sin \theta + b \cos \theta = 2q,$$

$$b \cos^3 \theta - a \sin^3 \theta = q \cos 2\theta.$$

Solving for q and θ gives

$$\tan \theta = \frac{b}{a}, \tag{2.16}$$

$$q = \frac{ab}{\sqrt{a^2 + b^2}}, \tag{2.17}$$

$$p = \sqrt{a^2 + b^2}. \tag{2.18}$$

Strictly from the calculus, we should now check that the conditions obtained do give the *minimum* A value and not the maximum. However, from reference to Fig. 2.8, it can be seen that A can increase without limit for small θ values; so we assume the solution in equations (2.16) and (2.17) corresponds to the required minimum.

Now can this be tested for its validity in giving the correct bed and corner dimensions? Suppose that a standard bed of length 2 m by width 1 m is tried. This means that $p = 2$ and $q = 1$. Calculating a and b from equations (2.17) and (2.18) gives $a = b = \sqrt{2}$ m.

On the other hand, what happens if the corridor is only 1.0 m across in one direction and 1.5 m across in the other? Taking $a = 1$ and $b = 1.5$, then from equations (2.17) and (2.18) we get $p = 1.8$ m and $q = 0.8$ m. This bed is then too short for a 6 ft person to lie in.

Follow-up

(a) Try other corridor sizes and bed sizes.
(b) The results can be validated in practice by cutting out a scale model of

a corridor system and corresponding bed. See whether you still agree with the above.

(c) Now consider the ladder again; suppose that it is tilted at an angle so that the height of the corridor comes into the problem. What difference does this make to the length of the longest ladder that can be carried round?

Example 2.2.10: Snowplough

Problem description

On the occasions when there is a sudden very heavy snowfall, country roads can easily become blocked. This necessitates the use of the local council's snowplough in order to clear a way through. In a particular situation a long 10 km stretch of road has to be cleared after a fall that results in an even depth of 0.5 m along the road. The clearance would be routine at a steady slow pace apart from the fact that, just as the snowplough is about to move off, it begins to snow again heavily. As the depth of snow increases, so the speed of the snowplough decreases until, at a certain depth, the snowplough itself can get stuck.

The rate of fall of the new snow will affect the progress of the snowplough and the question that arises is whether the snowplough can complete the 10 km clearance or find that the snow builds up so quickly that the snowplough gets stuck.

We shall consider the following data.

1 It snows for 1 h in total.
2 The snowfall rate may vary, but at its heaviest it is 0.1 cm s^{-1}.
3 The snowplough gets stuck when the depth of snow reaches 1.5 m.
4 On a road completely clear of snow, the plough travels at 10 m s^{-1}.

Procedure

To make a start, let us suppose that the speed v of the plough decreases in direct proportion to the depth d of snow. Measuring v in metres per second and d in metres, we arrive at the formula

$$v = 10\left(1 - \frac{2d}{3}\right) \tag{2.19}$$

valid for $d \leqslant 1.5$, after which the snowplough has stopped. Other relations can be constructed for calculating the speed in terms of the depth of snow, but this is the most simple. Note that this relation satisfies the data given: $v = 10$ when $d = 0$ and $v = 0$ when $d = 1.5$. Also using equation (2.19), since the initial depth of snow is 0.5 m, the plough moves off on its journey at 6.7 m s^{-1}.

Now the new snowfall rate has to be modelled by a suitable function. Earlier we said that the rate can vary throughout the hour of the fall. However, to make the model simple to begin with, suppose that the snowfall rate is taken as constant. Denote this by R_{const} (cm s^{-1}). Therefore the increase in depth of snow is $R_{\text{const}}t$ cm $= R_{\text{const}}t/100$ m in t s. The total snow depth now is

$$d = 0.5 + \frac{R_{\text{const}}t}{100}. \tag{2.20}$$

This enables the speed of the plough after t s to be obtained by substituting from equation (2.20) into equation (2.19), giving

$$v = \frac{10}{3}\left(2 - \frac{R_{\text{const}}t}{50}\right). \tag{2.21}$$

If v has declined to zero, then the plough has stopped, and this occurs when

$$t = \frac{100}{R_{\text{const}}}. \tag{2.22}$$

The distance travelled in this time by the snowplough will be obtained by integration:

$$\begin{aligned}\text{distance } s &= \int v \,\mathrm{d}t \\ &= \frac{10}{3}\int\left(2 - \frac{R_{\text{const}}t}{50}\right)\mathrm{d}t \\ &= \frac{20t}{3} - \frac{R_{\text{const}}t^2}{30}. \end{aligned} \tag{2.23}$$

(Remember that $s = 0$ initially.)

Now try some data to see what happens.

Case A Suppose that a heavy snowfall takes place at the maximum rate of 0.1 cm s^{-1}. If snow continued to fall at this rate for the entire hour, then the extra depth of $0.1 \times 3600/100$ m $= 3.6$ m would accrue on top of the original 0.5 m, which is very unlikely. However, before rejecting this forecast, we shall see what has happened to the snowplough by using equations (2.22) and (2.23).

From equation (2.22), the plough stops when $t = 100/0.1$ s $= 1000$ s $= 16.67$ min.

From equation (2.23), we can work out that the plough has travelled a distance of $20 \times 1000/3 - 0.1 \times (1000)^2/30$ m ≈ 3333.33 m $= 3.33$ km. So the plough travels exactly a third of the way along the road before getting stuck. The depth of snow at this moment is obviously 1.5 m, which is a long way short of the projected final depth of 4.1 m after 1 h. However, once the plough has stopped, it remains stuck until further relief is available; so it may as well now stop snowing!

Case B Suppose now that a less heavy snowfall takes place at a steady rate of 0.025 cm s^{-1}. This time, from equation (2.22), the plough comes to rest after $100/0.025 = 4000$ s ≈ 66.67 min, which of course is four times as long as in Case A. The distance travelled in that time is again worked out from equation (2.23) as $20 \times 4000/3 - 0.025 \times (4000)^2/30$ m $\approx 13\,333.33$ m $=$ 13.33 km. However, this is further than the allotted 10 km on the road to be cleared; so we conclude that the snowplough has achieved its target, assuming another plough has responsibility for the additional road. The actual clearance time can be worked out by substituting into equation (2.23) that $s = 10 \times 1000$ and $R_{\text{const}} = 0.025$. The resulting equation can be reduced to

$$0.0025t^2 - 20t + 30\,000 = 0.$$

The solution that we want from this is $t = 2000$ s ≈ 33.33 min.

The speed of the plough after 10 km can be calculated to be 3.33 m s^{-1} by substituting into equation (2.21).

Finally, we shall consider a situation where the new snowfall rate is not constant. Suppose that the rate increases steadily up to the maximum rate of 0.1 cm s^{-1} after 30 min before decreasing again. The snow rate profile is shown in Fig. 2.9.

Denoting the rate by $r(t)$ cm s^{-1}, then

$$r(t) = \begin{cases} \dfrac{0.1t}{1800}, & 0 \leqslant t \leqslant 1800, \\[2ex] 0.2 - \dfrac{0.1t}{1800}, & 1800 \leqslant t \leqslant 3600. \end{cases}$$

The calculation of the depth of snow follows by integrating the expression

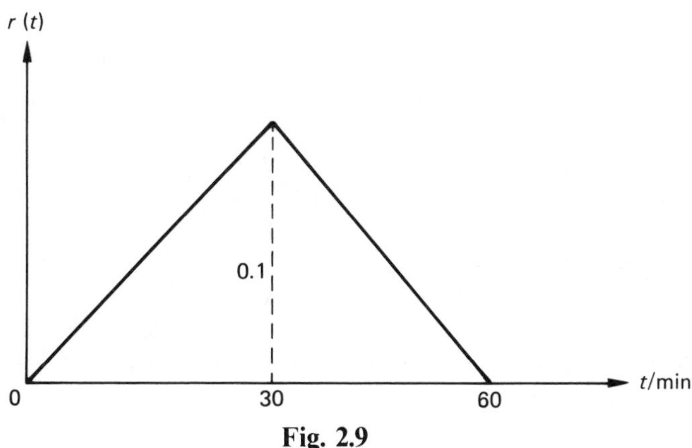

Fig. 2.9

for $r(t)$. The technique needs a little care owing to the nature of $r(t)$ where we have defined it by a different function over the separate time ranges.

For current time $T \leqslant 30$ min,

$$d = 0.5 + 0.01 \int_0^T \frac{0.1t}{1800}\,dt$$

$$= 0.5 + \frac{0.001T^2}{3600}. \qquad (2.24)$$

This gives a depth of snow of 1.4 m after $\frac{1}{2}$ h. For current time $T > 30$ min,

$$d = 1.4 + 0.01 \int_{1800}^T \left(0.2 - \frac{0.1t}{1800}\right) dt.$$

After a little simplification, this reduces to

$$d = 0.01\left(0.2T - \frac{0.1T^2}{3600}\right) - 1.3. \qquad (2.25)$$

At the end of an hour, the depth of uncleared snow can be calculated by substituting $T = 3600$ into equation (2.25). This gives a final uncleared depth of 2.3 m.

We must now return to equation (2.19) to obtain the expression for the speed of the plough. From this, the distance can then be evaluated by integration. You should attempt to complete this now to see whether the snowplough gets stuck or not. (The answer is given at the end of the chapter).

Follow-up

(a) Try other snow rate profiles.

(b) What is happening to the snow depth behind the plough? This is an interesting problem, especially with a variable snowfall rate. Immediately behind the plough the road is clear but, further behind, the snow is beginning to build up again. If the situation were photographed after, say, 20 min, what would the picture show?

(c) Develop alternative ways of formulating the model. For example, a reasonable assumption is that the plough works at a constant rate. This means that it removes snow at a constant rate. How does this alter the speed–distance relationship?

2.3 CONCLUSION

The 10 examples described above are there to give you the idea of what modelling is all about. In each case the level of mathematical technique required is deliberately kept at a moderate level so that the principles of modelling can be emphasised. It may be useful for chapters 5 and 6 to be read alongside the current chapter, since they contain important skills and ideas which can be used in developing the models.

2.4 FURTHER EXAMPLES

Example 2.4.1: Investment

When a sum of money is invested in a building society, it gains interest at a certain rate. Usually, for small savings accounts, you want to pay in money from time to time as and when you have money available. Also you will want to withdraw cash for a particular purpose—perhaps to pay for an annual holiday. This means that the calculation of the amount currently held in your account depends not only on the dates of deposits and withdrawals but also on how the interest is worked out. The interest rate often varies as well, just to complicate matters.

Formulate a mathematical model which will tell you how much is in your account at any given time. (Take the calculation of interest as compounded each month and decide, for example, that any deposits only gain interest from the start of the next month after being invested.)

Consider the 'reverse' problem of needing to save a certain amount by 12 months' time for particular use. How much must you invest each month?

Example 2.4.2: Ladders

There is a conservatory behind your house which sticks out into the garden by 7 ft. The conservatory roof is flat and is 10 ft above ground level. Above the conservatory at first- and second-floor level are bedroom windows to be cleaned. The conservatory runs for the whole width of the property.

The window cleaner attempts to place his ladder against the house wall, resting on the ground and passing over the conservatory. He does not have many ladders to choose from but starts with a ladder of 20 ft in length after extending it. Can this ladder reach over the conservatory and touch the house wall? If so, what height does it reach up the wall? What is the minimum ladder length that can just pass over the conservatory and still rest against the house?

Example 2.4.3: Art gallery security

A small art gallery has a security problem and decides to dispense with attendants on duty 'on the floor of the gallery' and instead to install a television eye system in which cameras pan over the area of the gallery. The resulting pictures are then watched from a remote control room. At the gallery entrance, an attendant checks tickets whilst at the exit another gives a cursory check of people as they leave. The current exhibition is that of a collection of small highly valuable water-colours. The layout of the gallery is shown in Fig. 2.10. The management decide to make do with two television cameras as shown.

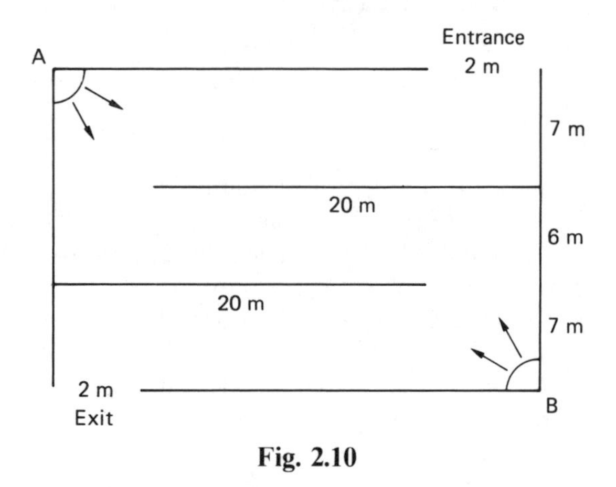

Fig. 2.10

There is a water-colour placed every metre around the walls of the gallery without any gaps. The television cameras give at any instant a 'scan beam' of 30°. They rotate backwards and forwards over the field of vision, taking 20 s to complete one cycle. In the control room the attendant looks at the television pictures 50% of the time since he has other jobs to do. Analyse the success of this system. How many water-colours are under surveillance at a particular instant? Calculate the relation between 'time' and 'number of water-colours' throughout one camera cycle. Suppose a thief tries to steal a water-colour—what are his chances of escaping television detection? What happens if the rotation speed of the cameras is changed?

Example 2.4.4: Pyramid selling

'Pyramid' selling of household articles through gatherings of buyers and sellers at people's houses is quite a well-known marketing technique. For example, a range of plastic tableware (plates, cereal bowls, etc.) are often sold in this way at so-called Tupperware parties. The principal features of pyramid selling are as follows.

The Tupperware agent in the district calls a party to which a number of people come. They pay a 'joining fee' on arrival. The Tupperware agent then demonstrates the qualities of the particular items currently on sale and sells some to those present. The Tupperware agent then invites each guest to become a subagent and to organise their own parties in their own houses to which they will invite new sets of people. When such further parties take place, the subagents behave similarly, i.e. they sell the tableware and tell each guest to become an agent. This process then continues over more 'levels' as appropriate, thus creating a pyramid of 'members'. The whole system can start again with a different product and be repeated.

Can you formulate the pyramid selling situation into a mathematical model? Remember that each subagent pays a certain percentage of profit back to the agent one level higher up in the pyramid.

Suppose, for simplicity, that one particular Tupperware article is sold. Suppose that each member buys the same quantity of goods.

Develop the model to deal with a variable number of articles sold. What happens if the number of people at the parties varies? How many 'levels' of the pyramid are feasible? What happens if some of (or all) the guests are agents from a higher level?

2.5 FURTHER COMMENTS ON THE EXAMPLES IN SECTION 2.2

Example 2.2.1

The tape example has a formula relating tape width to angle of pitch for a particular pipe. The relation can be obtained by reference to Fig. 2.11 and this gives the equation $W = C \sin A$. Points A′ coincide for a pipe of circular section.

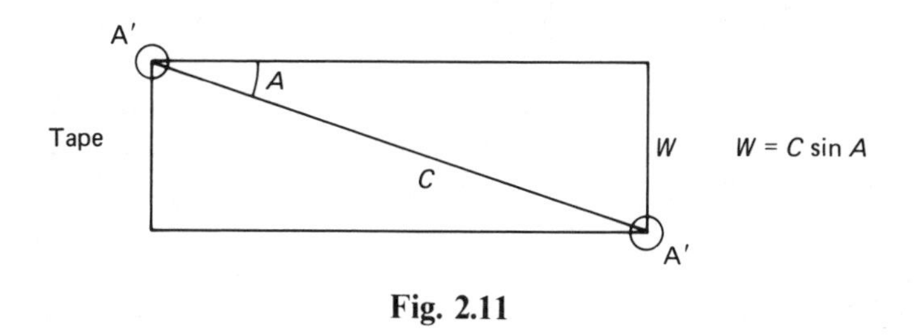

Fig. 2.11

Example 2.2.2

The fixtures for eight clubs should be as in Table 2.11.

Table 2.11

	Opponents for the following clubs							
Week	Club 1	Club 2	Club 3	Club 4	Club 5	Club 6	Club 7	Club 8
1	Club 2	Club 1	Club 7	Club 6	Club 8	Club 4	Club 3	Club 5
2	Club 3	Club 8	Club 1	Club 7	Club 6	Club 5	Club 4	Club 2
3	Club 4	Club 3	Club 2	Club 1	Club 7	Club 8	Club 5	Club 6
4	Club 5	Club 4	Club 8	Club 2	Club 1	Club 7	Club 6	Club 3
5	Club 6	Club 5	Club 4	Club 3	Club 2	Club 1	Club 8	Club 7
6	Club 7	Club 6	Club 5	Club 8	Club 3	Club 2	Club 1	Club 4
7	Club 8	Club 7	Club 6	Club 5	Club 4	Club 3	Club 2	Club 1

Example 2.2.5

Eight cars can certainly pass through the traffic lights. The ninth will probably also go across since it is travelling at about 31 miles h^{-1} and the stopping distance for the speed is 23 m.

Example 2.2.10

The snowplough results are as follows:

$$v = \begin{cases} 10\left(\dfrac{2}{3} - \dfrac{0.001\,T^2}{5400}\right), & 0 < T < 1800, \\ 10\left[1 - \dfrac{2}{3} \times 0.01\left(0.2T - \dfrac{0.1\,T^2}{3600}\right) - 1.3\right], & 1800 < T < 3600, \end{cases}$$

and

$$s = 10\left(\frac{2T}{3} - \frac{0.001\,T^3}{16\,200}\right), \qquad 0 < T < 1800.$$

This means that, after 30 min, the plough has travelled 8.4 km and its speed is 0.67 m s^{-1}. The snow is 1.4 m deep; so clearance continues, but not for long. Setting $v = 0$ gives the following quadratic equation in T:

$$0.1T^2 - 720T + 280 \times 3600 = 0,$$

which has a root at $T = 1903$, i.e. after another 103 s the plough gets stuck. The further distance gone can be calculated by integration to be about 34 m.

3 MODELLING METHODOLOGY

3.1 INTRODUCTION

A wide variety of models and problems have been described in the previous chapter. There are many alternatives that could have been included, and this indicates how large a total of real problems are available for investigation. Another reason for discussing such a mixture of examples was to raise a number of questions such as the following.

1 What is the difference between a 'model' and a 'problem'?
2 Is there some common thread running through all the examples shown in chapter 2?
3 Is there a general method that can be used to solve them?

In this chapter, we shall discuss some answers to these questions. The strict answer to the third question is clearly *no.* There are too many differences between individual problems for the same general method to be used for all of them. However, there is a general *approach* which can be used and this is the methodology which forms the title of this chapter. In order to make some progress towards answering questions 1 and 2, let us take another look at the examples in chapter 2. Some salient features from each example can be identified and are listed below.

Example 2.2.1: *Tape*
Specific problem; a unique formula required; an investigation; measurements taken; units chosen.

Example 2.2.2: *Fixtures*
Organisational; specific problem; not a unique answer; no measurements, units or formulation required; optimum sought.

Example 2.2.3: *Ferry*
Random effect; objective needs defining; one ferry load does not tell us much; need many 'runs' for statistical outcome.

Example 2.2.4: *Home decorating*
Tabular; specific (and simple) problem; clear objective.

Example 2.2.5: *Traffic lights*
Well-understood situation; observation and validation possible; units needed; car-following model produced.

Example 2.2.6: *Price war*
Relations need formulating; behavioural and strategic.

Example 2.2.7: *Evacuation*
Data needed; relations to be formulated; specific answer in this case.

Example 2.2.8: *Crossing the road*
Random effect; speculative; data needed; specific answer.

Example 2.2.9: *Corner*
Specific problem; formulation needed; validation possible.

Example 2.2.10: *Snowplough*
Relations need formulating; variables and units to be chosen; validation difficult; further development possible.

We can see from this list that some examples required particular answers to particular questions. In others, there is more than one answer, owing to either the nature of the example or the formulation presented. Also some seem relatively easy and 'closed' (e.g. Example 2.2.4), whereas others can be extended and are 'open' in the sense that we may want to vary our approach or develop more complicated relations between certain variables (as in Examples 2.2.2, 2.2.3, 2.2.6 and 2.2.10).

These considerations partly answer question 1; a problem often has a specific *correct* answer. A model is more general and speculative. Very often a model deals with quantities in a *general* way, represented by symbols without specifying particular values. Different models can be developed for the *same* situation and different answers can be obtained. It is not the case that one will be 'correct' and all the others 'wrong', although some may well be more useful than others. Solving a problem often requires insight and the use of an appropriate technique. Developing a model requires these qualities together with some creative imagination.

There are other issues which have emerged from the above list. In each example, certain *assumptions* may have been made concerning the relations between the variables in a problem. A choice of *units* was often necessary so that calculations could be done. The essential features of the problem were often *modelled* or *formulated* in mathematical terms and some conclusion or result obtained from the mathematics. We then try to carry out some *interpretation* or *validation* to see whether the mathematical results are worth anything and whether they stand up to *reality*. You may also have realised that, if the outcome is actually wanted by someone (and we hope that it is!), then the person will probably not be a mathematician. It will be the planner, engineer, salesman, local council, football league, etc.; so we shall have to *present* the results to a non-expert who may require a variety of detail—perhaps a long *report* or maybe just a short *verbal* statement. How to *present* and *report* on the findings of a modelling activity will be discussed in chapter 9.

To sum up by returning to the questions listed at the start of this section, although it is not possible to give clear crisp answers, we can begin to see what the important issues are. Hopefully, you will have appreciated the necessity to state the problem objective in each case. Then there is an ability to convert the problem into a mathematical form by identifying relations between the variables. This is one 'common thread' running through all the examples from chapter 2 and is one distinguishing feature in a modelling activity compared with the conventional use of mathematics. There is enough similarity in the manner that each example is tackled for a structured approach to modelling to be worth developing. This structure then provides an important guide when we are faced with new situations requiring a mathematical model. Such a 'modelling methodology' will be explained in section 3.3.

3.2 DEFINITIONS AND TERMINOLOGY

A number of different terms are commonly used to help to classify models and it will be useful to explain them here. The term *mathematical modelling* itself has been described in chapter 1; so there is no need to cover the same ground again. Perhaps a reminder is appropriate that the word 'mathematics' is used 'generically' to include statistics, operational research and computing as well as conventional mathematics.

The term *problem solving* was mentioned in chapter 2, suggesting a slightly different activity from modelling. Without being too rigid in our definitions, this term is often reserved for those situations where a definite well-defined 'problem' has to be tackled and a particular answer is required. Expressing the problem in mathematical terms will not lead to much argument over alternatives. Examples 2.2.1, 2.2.4 and 2.2.9 come into this category.

The term *simulation* can often be used synonymously with 'modelling', but some writers prefer to save it for use when the situation being modelled is very complex or more especially when there are random effects to be considered. Note in passing that *The Oxford English Dictionary* definition of 'to simulate' is 'to assume the mere appearance of', 'to feign', 'to counterfeit' or 'to pretend'. This is not what we want at all since it suggests some sort of confidence trick! To explain what *is* meant, consider the following two examples.

Example 3.2.1

A health centre is to be set up in a small town to contain, under one roof, a team of both doctors and pharmacists. The local area health authority want to know whether there is likely to be any improvement in efficiency of this new system compared with that currently in use, where doctors all act individually and pharmacists work in separate chemist's shops. In particular, the local authority need to judge the efficiency *before* going ahead with building work.

The operation of the new centre can be tested by constructing a model of the situation using statistics and mathematics. Data would have to be collected about doctors' current consultancy times, number of doctors and pharmacists to be available and so on. The data would then be used to construct a model that *simulates* the real system. Repeated operation of the model would be carried out to provide output on the times patients spent at the centre, the times a doctor was idle and so on. The results would be averaged so that items of information such as 'mean patient time at the centre' and 'mean number of patients treated per day' could be calculated. These results would then be compared with similar calculations for the current system so that a decision could be made. Essentially the proposed new system is being *simulated* instead of being put into operation and then tested, when it may be too late to reverse the decision.

Example 3.2.2

British Gas transport natural gas under high pressure around Britain prior to its local distribution. When North Sea natural gas was first available, they had the problem of guaranteeing to customers, both industrial and domestic, that sufficient gas would always be available. There was no available pipe network which could be used to transmit the gas. Another complication is that the demand for gas fluctuates with seasonal and daily weather conditions.

As in Example 3.2.1 the answers were needed *before* putting large resources into building the system. A mathematical model was formulated, based upon the known physical properties of gas and how it would flow along large, very long pipes. The model was then operated under various conditions to *simulate* the real system. A number of different pipe layouts were tried to obtain an optimal design. Again, only after many *simulations* had led to a confident prediction that gas demand could be met in all circumstances was the construction of the pipe network authorised.

In both the above examples it can be appreciated that, having formulated quite complicated *models*, the model would be translated into a *computer program* ready for subsequent *simulation*. The problems described in these two examples may seem far removed from the work of chapter 2 but, to describe the terminology adequately, it is necessary to raise our horizons a little.

The description *deterministic model* is usually used in cases where the outcome is a direct consequence of the initial conditions of the problem. This directness is not affected by any arbitrary external factors or, in particular, random factors. Very often, but not always, this kind of mathematical model involves differential equations in which time is the independent variable. In fact the gas flow model from British Gas mentioned above is one such deterministic model where differential equations are used. The gas flow received at some remote location will depend entirely on the initial state at a North Sea terminal and on the pipe network. Many models described in this book will be found to involve deterministic differential equations and chapter 8 is devoted to this particular topic.

The term *stochastic* model, on the other hand, is reserved for those situations where a *random* effect plays a central role in the problem investigation. We have already seen examples in chapter 2 (Examples 2.2.3 and 2.2.8). Many models of this kind are essentially 'next-event' models often involving queues and services. Random arrivals at bus queues or random service times at the supermarket are common events for everyone. In these situations the outcome is not fixed in the sense that it is unique because we have to allow for the random variability of arrivals and departures. This of course was the case in Example 3.2.1, where repeated simulations have to be statistically appraised before any sort of answer can be given. Stochastic models form the subject of chapter 7.

To sum up, it is often useful to divide our mathematical models into *two* categories, *deterministic* and *stochastic*, but care is necessary because there are many situations where, within the same model, some features are random and others deterministic.

3.3 METHODOLOGY AND MODELLING FLOW CHART

As has been said earlier, one of the main purposes of this *guide* is to teach how mathematical modelling is done in practice. One of the important conclusions from the previous section is that the activity of modelling is a *process* which involves a number of clearly identifiable *stages*. The most helpful way of representing these stages is by means of a modelling flow

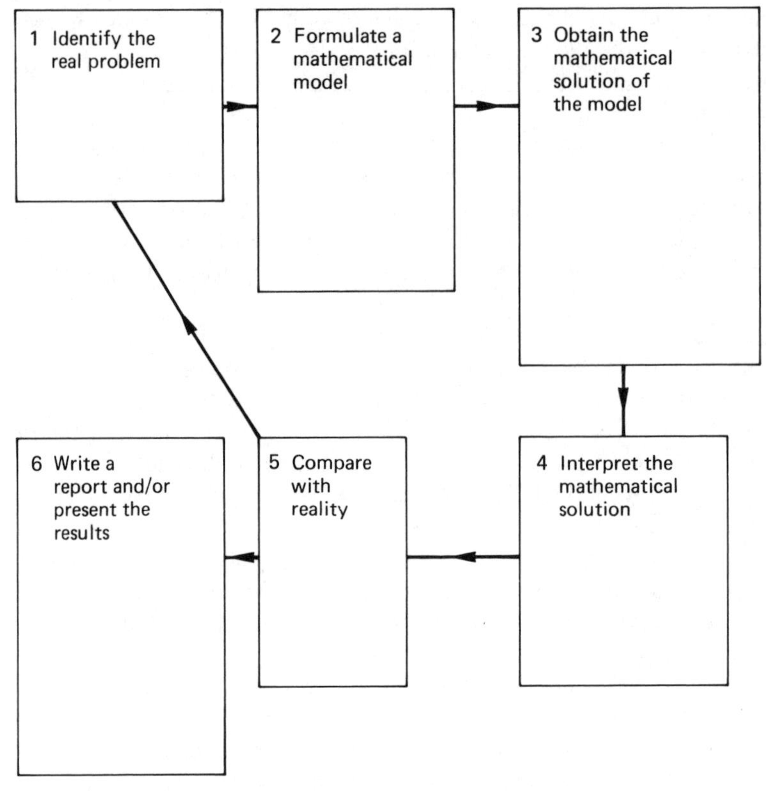

Fig. 3.1

chart. Our particular version of this flow chart is illustrated in Fig. 3.1. We do not claim complete originality for this chart and most books on mathematical modelling will have something similar. Experience shows that the flow chart does help in developing the right attitudes, leading to successful model building. There is perhaps some danger in trying to fit all situations onto the flow chart in a rigid manner but the comforting framework which it provides usually outweighs these dangers. The main point to remember is that when faced with a modelling problem, you should not be disconcerted if you feel 'lost' and wonder where to start. This is a perfectly normal reaction, even for experienced modellers. The point of the flow chart is that it gives us a framework to refer to and acts as a channel for our thoughts and ideas.

In this flow chart, each clear stage in the modelling process is represented by a box. We shall now amplify each 'box' with a series of questions and hints which should indicate what is intended.

Box 1: Identify the real problem

What do we want to know? What is the purpose and objective? How will the outcome be judged? What are the sources of facts and data, and are they reliable? Is there one particular unique answer to be found? Classify the problem: is it essentially deterministic, or stochastic? Do we need to use simulation?

Box 2: Formulate a mathematical model

Look first for the simplest model. Draw diagrams where appropriate. Identify and list the relevant factors. Collect data and examine them for information explaining the behaviour of the variables. Collect more data if necessary. Denote each variable by an appropriate symbol and assign units. State any assumptions that you decide to make. Draw up relations and equations connecting the problem variables, using your mathematical skills, e.g. proportionality, linear and non-linear relations, empirical relations, input–output principle, Newton's laws of motion, difference and differential equations, matrices, probability, statistical distributions, etc. (See chapter 5 for some help with this stage.)

Box 3: Obtain the mathematical solution of the model

Use algebraic and/or numerical methods, calculus and graphs. Write computer programs or use a prepared package if suitable. Use a simulation package if necessary (see section 7.6). Extract values for the variables that you want, either tabular or in graphical form.

Box 4: Interpret the mathematical solution

Examine the results obtained from the mathematics. Have the values of the variables got the correct sign and size? Do they increase or decrease when they should? Should a certain graph be linear? Consider large and small values of the variables to check for sensible behaviour. Have you got the 'best' solution that you expected or should some initial conditions be changed? (See chapter 6 for some help with this stage.)

Box 5: Compare with reality

Can your results be tested against real data? Do your mathematical solutions make sense? Do your predictions agree with the real data? Evaluate your model. Has it fulfilled its purpose? Can the model be significantly improved by greater mathematical sophistication? Do the interim results suggest that more accuracy is needed by rerunning with an improved model? *If yes then go to Box 1; otherwise go to Box 6.* This is important; very often the 'modelling cycle' is traversed a number of times before the results are satisfactory.

Box 6: Write a report

Who is the report for and what do the readers want to know? How much detail is required in the report? How can we construct the report so that the important features are clear and the results that we want to be read stand out? (See chapter 9 for more details.)

It is especially important to get off to a good start; so particular care should be taken with the first step, identifying the real problem. Think clearly; try to get to the heart of the problem. What is *given*? What are you asked to *find*? Do not go any further until you get these clear.

You may find the above methodology somewhat cumbersome and it is not necessary to keep to this format too rigidly. We want to be flexible in approach; so the structure described is a guide to good modelling practice. In developing models in subsequent chapters, this structure has sometimes been replaced by a simpler list of headings including the following.

Context	(Where does the problem come from and where does it fit in?)
Problem statement	(Define as clearly and completely as possible.)
Objective	(Condense even further: state exactly what is given, and exactly what you have to find.)
Model	(Formulate a mathematical model and interpret the solution.)

You should now return to the examples in chapter 2 and see how they fit in with the methodology described. A new modelling example will be given in the next section to illustrate how the modelling flow chart works.

3.4 THE METHODOLOGY IN PRACTICE

We now need an example that demonstrates the methodology, showing how a structured approach can benefit the modelling process. For this purpose, we shall take a well-known situation of common experience.

Background to the problem

It is about to rain; you have to walk a short distance of about 1 km between home and college. As there is some hurry, you do not bother to take a raincoat or umbrella but decide to 'chance it'. Suppose that it now starts to rain heavily and you do not turn back; how wet will you get?

This seems a simple matter of getting out of the rain as soon as possible but, if variation in the direction of the rainfall is taken into account, it may not follow that the best strategy is to run as fast as possible over the distance.

We shall now attempt to model this problem with the help of the flow chart structure.

Box 1: Identify the real problem

Given particular rainfall conditions, can a strategy be devised so that the amount of rain falling on you is minimised? The model will be 'deterministic' since it will depend entirely on the input factors such as the following.

(a) How fast is it raining?
(b) What is the wind direction?
(c) How far is the journey and how fast can you run?

We shall need to develop a formula for the amount of rain collected which is dependent on these factors. Suppose that the data available are as follows:

$$\text{walking speed} = 2 \text{ m s}^{-1},$$

$$\text{running speed} = 6 \text{ m s}^{-1},$$

$$\text{journey distance} = 1 \text{ km} = 1000 \text{ m},$$

$$\text{rainfall speed} = 4 \text{ m s}^{-1},$$

$$\text{rainfall intensity} = 2 \text{ cm h}^{-1}.$$

These data are typical for average behaviour but could be altered to cover more extreme cases.

Box 2: Formulate a mathematical model

The first objective is to set up the simplest model possible.

Suppose that you run the whole kilometre journey at 6 m s^{-1}. Therefore,

$$\begin{aligned}\text{time spent in the rain} &= \frac{1000}{6} \text{ s}\\ &\approx 167 \text{ s}\\ &= 2 \text{ min } 47 \text{ s}.\end{aligned}$$

Now ignore the rain direction, and merely consider rain collection from the given data of 2 cm h^{-1}, i.e. 2/3600 cm s^{-1}. Thus, over the whole journey of 167 s,

$$\begin{aligned}\text{amount of rain collected} &= \frac{2 \times 167}{3600} \text{ cm}\\ &= \frac{2 \times 167 \times 0.01}{3600} \text{ m}.\end{aligned}$$

It is now necessary to give some data about the surface area of the body which is being rained on. Suppose for simplicity that the human frame is represented as a rectangular block 1.5 m high, 0.5 m across and 0.2 m deep. Then

$$\begin{aligned}\text{front and back surface area} &= 1.5 \times 0.5 \times 2\\ &= 1.5\ \text{m}^2,\\ \text{sides surface areas} &= 1.5 \times 0.2 \times 2\\ &= 0.6\ \text{m}^2,\\ \text{top surface area} &= 0.5 \times 0.2\\ &= 0.1\ \text{m}^2,\end{aligned}$$

i.e.

$$\text{total surface area} = 2.2\ \text{m}^2.$$

On the assumption that all these surfaces collect rain, then

$$\begin{aligned}\text{volume collected} &= \frac{2 \times 167 \times 0.01 \times 2.2}{3600}\ \text{m}^3\\ &\approx 2.041 \times 10^3\ \text{cm}^3\\ &= 2.041\ \text{l}. \qquad (3.1)\end{aligned}$$

(So it is like having about two bottles of wine poured over you!)

The rules have been broken somewhat here by rushing through a quick result for illustration purposes, and achieving an answer that is reasonably plausible as well. It will be useful now to go back to the flow chart and to proceed through in a more detailed fashion incorporating this time all the relevant features, thus developing a second and more general model.

Assumptions

It has already been decided that the human frame can be represented by a rectangular block. A diagram helps in explaining the situation to be modelled and this is shown in Fig. 3.2. Other assumptions which we shall make are

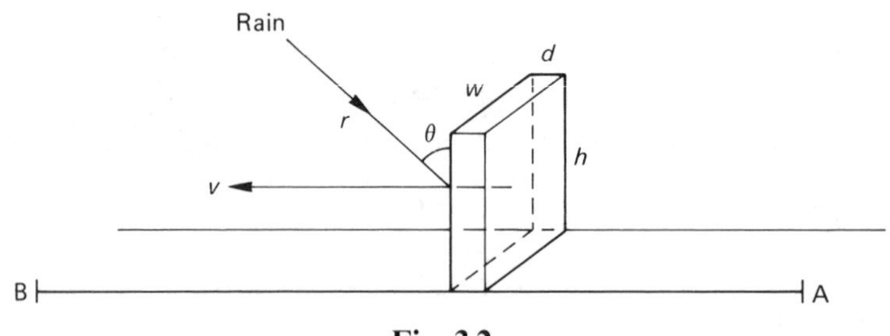

Fig. 3.2

that the rain speed is constant throughout and also that you move through the rain at a constant speed.

List of the factors

Description	Symbol	Units
Time during path through rain	t	s
Velocity of rain	r	$m\ s^{-1}$
Angle of rainfall (due to wind)	θ	deg
Your velocity	v	$m\ s^{-1}$
Personal dimensions		
Height	h	m
Width	w	m
Depth	d	m
Collection of rainwater on clothes	C	l
Rain intensity factor	I	—
Distance travelled	D	m

Some of these quantities are not variables at all but have numerical values from the data provided. It is nevertheless convenient to retain symbols while the model is being constructed. In fact, r, θ, v, t and C are variables while the others are 'parameters' in the sense that they do not vary for this particular situation. There is a need to distinguish between the *velocity* of rain and the *collection* of rain. If rain was a continuous flow of water (like a river), then the velocity of the rain would give us the collection rate over a certain area. However, this is clearly incorrect since rain is a stream of droplets which gives rise to the idea of rain intensity.

A rain intensity factor I is introduced to deal with this situation. From the data given above, the rain speed is $4\ m\ s^{-1}$ and the rain collection is taken as $2\ cm\ h^{-1}$. However,

$$\begin{aligned} \text{rain speed} &= 4\ \text{m s}^{-1} \\ &= 400 \times 3600\ \text{cm h}^{-1} \\ &= 1.44 \times 10^{6}\ \text{cm h}^{-1} \end{aligned} \tag{3.2}$$

compared with the collection rate of $2\ cm\ h^{-1}$, i.e. the ratio is 7.2×10^5. This discrepancy is allowed for by introducing I as the measure of rain intensity. For these data, $I = 1/(7.2 \times 10^5)$. Thus generally, if $I = 0$, it has stopped raining, while increasing I gives heavier rain. Ultimately, $I = 1.0$ would correspond to continuous flow like a river.

We are now ready to draw up the equations relating the variables listed above. There is no question of the laws of motion, probability effects, etc.,

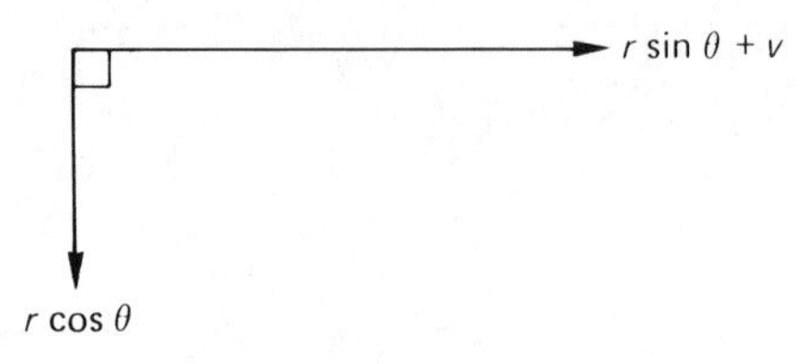

Fig. 3.3

to worry about, merely an evaluation of the rain collection capacity. With speed taken as constant, then

$$\text{time spent in the rain} = \frac{D}{v} \quad (\text{s}). \tag{3.3}$$

Also the key factor now to be considered in assessing how wet you get is the *relative* direction of the rain with respect to your direction of travel. These relative effects are conveniently shown in Fig. 3.3.

Now, since the rain is coming down at an angle, we can see that in any situation only your *top* and *front* will be getting wet. This is in accord with experience.

Box 3: Obtain the mathematical solution

The amount of rain falling on you will now be calculated in the two cases.

First, consider your top surface area, for which

$$\text{top area being rained on} = wd \quad (\text{m}^2)$$

and

$$\text{rain component (Fig. 3.3)} = r \cos \theta.$$

Therefore,

$$\begin{aligned}\text{rate of rain collection} &= \text{intensity} \times \text{area} \times \text{rain speed}\\ &= Iwdr \cos \theta \quad (\text{m}^3\ \text{s}^{-1}).\end{aligned}$$

In time D/v,

$$\text{amount of rain collected} = \frac{IwdDr \cos \theta}{v} \quad (\text{m}^3). \tag{3.4}$$

Next consider your front surface area, for which

$$\text{front area being rained on} = wh \quad (\text{m}^2),$$

and

$$\text{rain component (Fig. 3.3)} = r \sin \theta + v.$$

Therefore,

$$\text{rate of rain collection} = \text{intensity} \times \text{area} \times \text{rain speed}$$
$$= Iwh(r \sin\theta + v) \qquad (\mathrm{m}^3\ \mathrm{s}^{-1}).$$

In time D/v,

$$\text{amount of rain collected} = \frac{IwhD(r\sin\theta + v)}{v} \qquad (\mathrm{m}^3). \qquad (3.5)$$

By addition of equations (3.4) and (3.5), the total amount C of rain collected is

$$C = \frac{IwD}{v}[rd\cos\theta + h(r\sin\theta + v)] \qquad (\mathrm{m}^3). \qquad (3.6)$$

It is now a question of extracting from equation (3.6) the information that we want on how wet the journey will be. First, values for some of the quantities used can be substituted. From the earlier data, we have $h = 1.5$, $w = 0.5$ and $d = 0.2$; also $r = 4$ and $D = 1000.0$. Also substituting the value for I from equation (3.2), then

$$C = \frac{0.8\cos\theta + 6\sin\theta + 1.5v}{1.44 \times 10^3 v} \qquad (\mathrm{m}^3). \qquad (3.7)$$

The variables retained are v and θ, since you can choose v and θ is the rain direction, which we shall want to vary in investigating the mathematical solution. Thus, given θ, what v is chosen so that C is minimised?

Box 4: Interpret the mathematical solution

Equations (3.6) and (3.7) are now examined. First, note that, if the rain intensity I is zero, then $C = 0$ which means that you stay dry. Second, the value of θ will determine whether the rain is facing you or blowing in from behind. We shall evaluate equation (3.7) to show what happens in particular cases.

Case 1: $\theta = 0°$

In this case, as $\theta = 0°$, the rain is falling straight down. From equation (3.7),

$$C = \frac{0.8 + 1.5v}{1.44 \times 10^3 v} \qquad (\mathrm{m}^3).$$

This expression is smallest for the largest possible value of v, i.e. $v = 6$ in this case.

Substituting $v = 6$ gives

$$C = \frac{9.8}{1.44 \times 10^3 \times 6} \qquad (\mathrm{m}^3)$$
$$\approx 1.13\ \mathrm{l}. \qquad (3.8)$$

Case 2: $\theta = 30°$

In this case, as $\theta = 30°$, the rain is driving in towards you. From equation (3.7),

$$C = \frac{(0.4\sqrt{3} + 3 + 1.5v)}{1.44 \times 10^3 v} \qquad (\text{m}^3).$$

Again this is at its smallest when $v = 6$, i.e.

$$C = \frac{0.4\sqrt{3} + 3 + 9}{1.44 \times 6} \qquad (\text{litres})$$

$$= 1.47 \text{ litres.} \tag{3.9}$$

Case 3: Negative θ

Now suppose that the rain is coming from the rear so that θ is negative. Taking $\theta = -\alpha$, say, then from equation (3.7), we obtain

$$C = \frac{(0.8 \cos \alpha - 6 \sin \alpha + 1.5v)}{1.44 \times 10^3 v} \qquad (\text{m}^3).$$

This expression can become negative for α sufficiently large, which means that the model must be examined more carefully since it is not possible for C to be negative! It is best to return to equation (3.5) to analyse the situation. With $\theta = -\alpha$, there are two cases to consider according to how fast you move through the rain.

In the first case, for $v < r \sin \alpha$, it rains on your *back* and so

$$\text{amount of rain collected} = \frac{IwD}{v}(r \sin \alpha - v).$$

This gives the total collection formula now as

$$C = \frac{IwD}{v}[rd \cos \alpha + h(r \sin \alpha - v)].$$

Putting in the data again, we get

$$C = \frac{0.8 \cos \alpha + 1.5(4 \sin \alpha - v)}{1.44 \times 10^3 v} \qquad (\text{m}^3). \tag{3.10}$$

If you now increase your speed to $4 \sin \alpha$, this expression is reduced to

$$\frac{0.8 \cos \alpha}{1.44 \times 10^3 \times 4 \sin \alpha}$$

which corresponds to the *top* rain amount only. Thus, if the rain is at an angle of 30° from the rear, you should walk at $4 \sin 30° = 2 \text{ m s}^{-1}$, in which case

$$\text{amount of rain collected} = \frac{0.8\sqrt{3}}{2.88 \times 10^3 \times 2} \quad \text{m}^3$$

$$= 0.24 \text{ litre.} \qquad (3.11)$$

Effectively this means that you are just keeping up with the rain. If the speed falls below 2 m s^{-1}, then the rain collected increases due to that falling on your back.

In the second case, should you be unable to resist the temptation to run at a speed of more than 2 m s^{-1}, then you will be catching up with the rain. This is the case where $v > r \sin \alpha$ and the contribution from equation (3.5) is now

$$\frac{IwhD}{v}(v - r \sin \alpha).$$

So the total amount of rain now collected is

$$C = \frac{IwD}{v}[rd \cos \alpha + h(v - r \sin \alpha)] \qquad (\text{m}^3).$$

When $v = 6$ and $\alpha = 30$,

$$C = \frac{0.5 \times 1000(0.4\sqrt{3} + 6)}{7.2 \times 10^5 \times 6} \qquad (\text{m}^3)$$

$$\approx 0.77 \text{ litre.} \qquad (3.12)$$

Box 5: Compare with reality

The results seem sensible and agree with what we might expect. In what way has this second detailed model improved on the earlier? Here we have allowed for wind direction and investigated the various cases more thoroughly. All the results for rain collection are less than the value of 2 litres obtained for the initial model. Also the orders of magnitude are what we might expect. It is difficult to validate the numerical results of the model, but the idea of 'moving with the rain' can be tried out in practice, assuming that you do not mind getting wet. The overall conclusions from the model are as follows.

1 If the rain is driving towards you, then the strategy should be simply to run as fast as possible.
2 If the rain is being blown from behind, then you should keep pace with the rain, which means moving with a speed equal to the wind speed.

Note in passing here that the conclusions are given in simple everyday terms that are easily understood. It is no good telling a non-mathematician to run at $r \sin \alpha$ (m s^{-1}).

Box 6: Write a report and present the results

We shall look at report writing and presentations in some detail in chapter 9. A sample report is given there and it may then be beneficial to return to the rain problem to write up your account.

3.5 SUMMARY

1 We have set out a structure for use in modelling assignments. The methodology will act as a guide to help to make a start when considering a particular newly presented situation that seems intractable. You must not become too encumbered by formal methods, however, since half the fun in creating mathematical models is to try your own ideas. However, a balance is needed between rushing in perhaps without adequate thought and being overcautious in trying to fit a rigid methodology to every situation.

2 The important issues for a successful approach to mathematical modelling have been systematically set out in section 3.3. It is advisable to refer to the items listed as you gradually gain in confidence and experience.

3 For future reference, it is convenient to have a clear understanding of the terms used in a modelling context. The brief discussion here in section 3.2 should help you to place the subsequent models treated in the book in the correct context.

4 UNITS AND DIMENSIONS

4.1 INTRODUCTION

As we have seen, mathematical modelling involves variables, parameters and constants, all of which represent quantities which can (in principle at least) be measured. Any physical quantity such as length or mass is measured by writing a number followed by a unit of measurement, e.g. a length of 2.6 m, where the m stands for metre. In this case the metre is the unit of measurement. A measurement without units is totally meaningless and we must be careful to keep track of the units of measurement for all quantities involved at all stages in the model-building process.

4.2 UNITS

There are several different systems of units which have been used in the past and unfortunately are still in use today. We therefore need to be clear about which particular system we are using and keep to it consistently. It will often be necessary to convert from one system to another. We may find that the data we need are in the 'wrong' units and we need to know the relevant conversion factors. Most of the examples in chapter 2 involved units (Examples 2.2.1, 2.2.3, 2.2.6, 2.2.8 and 2.2.10) and some (Examples 2.2.4 and 2.2.5) involved a mixture of units.

The recommended scientific system of units is the SI system (short for Système International d'Unités). This has seven basic units with accepted symbols as follows.

Quantity	Unit	Symbol
Length	metre	m
Mass	kilogram	kg
Time	second	s
Electric current	ampere	A
Temperature	kelvin	K
Luminous intensity	candela	cd
Amount of substance	mole	mol

There are also other commonly used units which are combinations of some of these and have been given their own names and symbols.

Quantity	Unit	Symbol
Force	newton	N (kg m s^{-2})
Energy	joule	J (kg m^2 s^{-2})
Power	watt	W (J s^{-1} or kg m^2 s^{-3})
Frequency	hertz	Hz (s^{-1})
Pressure	pascal	Pa (N m^{-2} or kg m^{-1} s^{-2})

Note the use of both positive and negative indices where a combination of units is involved. Speed, for example, is measured in metres per second which in this book is abbreviated to m s^{-1} (the alternative m/s is sometimes used in other work). Similarly, for acceleration (which is measured in metres per second per second), we use m s^{-2} rather than m/s^2 (m/s/s should not be used at all).

There are a number of non-SI units which are in common use by scientists and engineers. The main ones are as shown opposite.

In the real world, there is such a wide range of sizes for all sorts of quantities, it is inevitable that the seven basic SI units are inadequate in their basic form. So we use special symbols for large and small multiples of the basic units. These are based on powers of 10 and the names and symbols shown opposite are the prefixes most commonly used.

For example, the power output of a generator will often be measured in megawatts (MW), building measurements are often in millimetres (mm) and computer operating times are in nanoseconds (ns).

Angles are measured in radians (symbol, rad). 1 rad is the angle between two radii of a circle which cut off on the circumference an arc equal to the radius. In three dimensions, solid angles are measured in steradians (symbol, sr). Other units for angles are the degree (symbol, °) which equals $\pi/180$ rad,

Quantity	Unit	Symbol
Area	hectare	ha $(=10^4\ m^2)$
Volume	litre	l $(=10^{-3}\ m^3)$
Volume	millilitre	ml $(=10^{-6}\ m^3)$
Temperature	degree Celsius	°C (0°C $\approx$ 273 K)
Mass	gram	g $(=10^{-3}$ kg)
Mass	tonne	t $(=10^3$ kg)
Energy	kilowatt hour	kW h $(=3.6\times10^6$ J)
Energy	electronvolt	eV $(\approx 1.6\times10^{-19}$ J)
Energy	calorie	cal (= 4.1868 J)
Pressure	bar	bar $(=10^5$ Pa)
Pressure	atmosphere	atm $(\approx 1.013\times10^5$ Pa)

Multiplication factor	Prefix	Symbol
10^{12}	tera	T
10^9	giga	G
10^6	mega	M
10^3	kilo	k
10^{-2}	centi	c
10^{-3}	milli	m
10^{-6}	micro	μ
10^{-9}	nano	n

the minute of arc (symbol, ′) which is $(1/60)^\circ$ or $\pi/10\,800$ rad, and the second of arc (symbol, ″) which is $(1/60)'$ or $\pi/648\,000$ rad. Conversely, 1 rad $\approx 57.295^\circ$.

Another system of units which is still in use is the British or foot–pound–second (fps) system. The main units and their conversion factors are given in Table 4.1. (Use reciprocals to convert the other way.)

The continuing use of the Fahrenheit scale of temperature can be a nuisance. To convert from °F to °C use °C = 5(°F − 32)/9; conversely, °F = 9 × °C/5 + 32.

Note that the above table shows the recommended usage regarding abbreviations and symbols but it would be misleading to pretend that these are strictly adhered to in real life. In modelling, we can expect to come across data measured in a variety of units often mixed up together as well as different abbreviations and symbols for the same units. It is common, for example, to see in and ins for inches and s or sec or secs for seconds. ISO recommendations have been used in this book, i.e. in for inches, and s for seconds.

Table 4.1

fps unit	SI equivalent
inch (in)	0.0254 m
foot (ft)	0.3048 m
mile (5280 ft) (5 miles $\approx$ 8 km)	$1.609\,344 \times 10^3$ m
nautical mile (6080 ft)	$1.853\,184 \times 10^3$ m
acre	$4.046\,856 \times 10^3$ m^2 ($\approx$ 0.4 ha)
square foot (ft^2)	$9.290\,304 \times 10^{-2}$ m^2
cubic foot (ft^3)	$2.831\,685 \times 10^{-2}$ m^3
fluid ounce (fl oz)	$2.841\,306 \times 10^{-5}$ m^3
pint (pt)	$5.682\,613 \times 10^{-4}$ m^3
gallon (gal)	$4.546\,09 \times 10^{-3}$ m^3
ounce (oz)	$2.834\,952 \times 10^{-2}$ kg
pound (lb)	0.453 592 37 kg
hundredweight (cwt)	50.802 345 kg
ton (2240 lb)	$1.016\,047 \times 10^3$ kg
pound per cubic foot (lb ft^{-3})	16.018 463 kg m^{-3}
mile per hour (mile h^{-1} (sometimes mph))	0.447 04 m s^{-1}
pound-force per square inch (lbf in^{-2} (sometimes psi))	$6.894\,757 \times 10^3$ Pa
British thermal unit (Btu)	$1.055\,06 \times 10^3$ J
therm (10^5 Btu)	$1.055\,06 \times 10^8$ J
horsepower (hp)	7.457×10^2 W

Example 4.2.1

A girl who has a mass of 8 stones eats a 50 g bar of chocolate. The energy content of the chocolate is 4700 kcal kg^{-1}. Assuming that her digestive system is able to convert all the chocolate bar's energy into usable mechanical energy, would the chocolate give her the energy required to climb up a 1000 ft hill?

Solution

First let us calculate the energy content of the chocolate. This is $0.05 \times 4700 \times 4.1868$ kJ $\approx$ 984 kJ. Note that the energy content of foods are usually measured in kilocalories but the prefix kilo is very often dropped and the units are loosely referred to as 'Calories'. The average person's daily intake is from 2000 to 4000 kcal. The energy used up in climbing the hill is not easy to assess. Some will be used in overcoming friction (converted into heat energy) but we can assume that the largest part will go into the increase in gravitational potential energy which is given by mass $\times\, g \times$ height $=$

$(8 \times 14 \times 0.4536) \times 9.807 \times (1000 \times 0.3048)$ J ≈ 151.9 kJ. We conclude that the chocolate gives ample energy for the climb.

Financial conversions

Unlike the physical units that we have so far discussed, monetary conversion rates are not fixed and, when we need to convert from £ sterling to dollars, for example, we have to use the current exchange rate. Occasionally, in modelling, we do need to change units from one currency to another. Some examples from *The Times*, September 1987, are given below.

Name	Symbol	Conversion factor
pound sterling	£	1
US dollar	$	1.6430
Deutschmark	DM	2.9796
French franc	FFr	9.9319
Swiss franc	SwFr	2.4686
Yen	Y	236.34

EXERCISES 4.2

1 Petrol costs £1.73 gal^{-1} in Britain. Assuming that it is sold in West Germany at the same price, calculate the cost in deutschmark per litre (DM l^{-1}).

2 A man is walking along a road at a speed of 4 miles h^{-1}. Calculate his speed in metres per second (m s^{-1}).

3 Alternative units of force which are sometimes used are the poundal (symbol, pdl) (defined as the force which gives a 1 lb mass an acceleration of 1 ft s^{-2}) and the dyne (symbol, dyn) (defined as the force which gives a 1 g mass an acceleration of 1 cm s^{-2}). Calculate the number of poundals equivalent to 1 N and the number of dynes equivalent to 1 pdl.

4 In a high-pressure gas flow, the flow rate is sometimes quoted in 'millions of cubic feet per day' and at other times in 'cubic metres per second'. Calculate the conversion factors.

5 Which is the cheapest source of energy, milk at 700 cal kg^{-1} costing 25p pint^{-1} or petrol at 18 600 Btu lb^{-1} costing £1.73 gal^{-1}?

6 One athletics track has a perimeter of 400 m while another has a perimeter of 440 yd. Which is longer?

7 Calculate the Earth's angular speed in radians per second (rad s^{-1}).

4.3 DIMENSIONS

In mechanics, all quantities can be expressed in terms of the fundamental quantities mass, length and time, denoted by the symbols M, L and T. Any other physical quantity will be a combination of these three and the particular combination is referred to as the 'dimensions' of that physical quantity. For example, the dimensions of area are L^2 and the dimensions of density are ML^{-3} (or M/L^3). Note that dimensions are independent of the units used. For example, speed has dimensions LT^{-1} (or L/T) but could be measured in miles per hour or metres per second. It is convenient to use square brackets [ ] to denote 'the dimensions of...' so that

$$[\text{area}] = L^2,$$

$$[\text{speed}] = LT^{-1}$$

$$[\text{density}] = ML^{-3}.$$

Note that some quantities are dimensionless, in other words pure numbers, e.g. $[\text{angle}] = LL^{-1} = L^0$. From Newton's second law of motion, force = mass × acceleration; so the dimensions of force are MLT^{-2} and the (abbreviated) SI units for force are kg m s^{-2} (defined to be 1 N—see section 4.2). In ordinary conversation, it is very common to use 'mass' and 'weight' as equivalent terms but do *not* confuse the two. The weight of an object is the force exerted on it by the Earth's gravitational field. It is an experimental fact that, near the Earth's surface, all bodies in free fall accelerate at the constant rate of $g \approx 9.806\,65$ m s^{-2} or 32.174 ft s^{-2}. Consequently the weight of a mass of m kg is mg N. It may be helpful to think of a force of 1 N as the weight of a fairly large apple (mass of about 0.1 kg).

Any sensible equation must be dimensionally consistent, i.e. [left-hand side] = [right-hand side]. It is a good idea to carry out this check on all the equations appearing in a model. Modelling errors can often reveal themselves in this way. Note that any constants appearing in our equations can be dimensionless (i.e. pure numbers) or can have dimensions. For example, suppose that we are modelling the force on a moving object due to air resistance. If we assume the magnitude of the force F is proportional to the square of the speed v, this leads to a model of the form $F = kv^2$. Checking the dimensions of this equation, we have $[F] = [kv^2]$, i.e.

$$MLT^{-2} = [k][LT^{-1}]^2 = [k]L^2T^{-2}.$$

For consistency, we require $[k] = ML^{-1}$ and k will be measured in kg m^{-1}.

Note that if expressions involving $\exp(at)$ or $\sin(at)$ appear in our model, where t stands for time, the parameter a must have dimensions T^{-1} so that at is a dimensionless number.

If an equation involves a derivative, the dimensions of the derivative are given by the ratio of the dimensions.

For example, if p is the pressure in a fluid at any point, the pressure gradient in the direction of z is dp/dz and

$$\left[\frac{dp}{dz}\right] = \frac{[p]}{[z]} = \frac{ML^{-1}T^{-2}}{L} = ML^{-2}T^{-2}.$$

Similarly, for partial derivatives,

$$\left[\frac{\partial p}{\partial t}\right] = \frac{[p]}{[t]} = \frac{ML^{-1}T^{-2}}{T} = ML^{-1}T^{-3}$$

and

$$\left[\frac{\partial^2 v}{\partial x^2}\right] = \frac{[v]}{[x]^2} = \frac{LT^{-1}}{L^2} = L^{-1}T^{-1}$$

4.4 DIMENSIONAL ANALYSIS

This is a method of using the fact that the dimensions of each term in an equation must be the same to suggest a relationship between the physical quantities involved. This will not give the exact form of a function but is still useful. Suppose that we are trying to develop a model which will predict the period of a swinging pendulum. A list of the factors involved might include the length l, the mass m, the acceleration g due to gravity, the amplitude θ, air resistance R and the rotation of the Earth. If we restrict this list to the first four factors, we can assume that the period t is given by some function of the four factors. Assume that $t = kl^a m^b g^c \theta^d$, where a, b, c, d and k are real numbers. Considering dimensions, we have

$$[t] = [kl^a m^b g^c \theta^d]$$

and so

$$T = L^a M^b (LT^{-2})^c$$

(k and θ are dimensionless). Equating powers of M, L and T on both sides, we must have

$$a + c = 0, \qquad b = 0 \qquad \text{and} \qquad -2c = 1.$$

This gives

$$t = kl^{1/2} g^{-1/2} \theta^d.$$

In this expression, d could take any value and we could sum terms of this form to arrive at $t = f(\theta)l^{1/2}g^{-1/2}$. We have to find $f(\theta)$ some other way. For small θ, of course, $f(\theta)$ is approximately constant with value 2π.

If we also include the force R due to air resistance, our model becomes

$$t = kl^a m^b g^c \theta^d R^e,$$

and so

$$\begin{aligned} M^0L^0T^1 &= L^a M^b (LT^{-2})^c (MLT^{-2})^e \\ &= M^{b+e} L^{a+c+e} T^{-2c-2e}. \end{aligned}$$

We require

$$b + e = 0,$$

$$a + c + e = 0$$

and

$$-2c - 2e = 1, \qquad \text{any } d.$$

We now have three equations with four unknowns. We could write any three of them in terms of the fourth. Suppose that we write everything in terms of b. We have

$$\begin{aligned} e &= -b, \\ c &= -b - \tfrac{1}{2}, \\ a &= \tfrac{1}{2}. \end{aligned}$$

So

$$\begin{aligned} t &= kl^{1/2} m^b g^{b-1/2} R^{-b} \theta^d \\ &= k\left(\frac{l}{g}\right)^{1/2} \left(\frac{mg}{R}\right)^b \theta^d, \end{aligned}$$

which generalises to

$$t = \left(\frac{l}{g}\right)^{1/2} f_1\left(\theta, \frac{mg}{R}\right)$$

for some function f_1.

Returning to the three equations with four unknowns, if we express all the parameters in terms of c, we have

$$\begin{aligned} e &= -c - \tfrac{1}{2}, \\ a &= \tfrac{1}{2}, \\ b &= c + \tfrac{1}{2}, \end{aligned}$$

giving

$$t = kl^{1/2} m^{c+1/2} g^c \theta^d R^{-c-1/2}$$

or

$$\left(\frac{ml}{R}\right)^{1/2} f_2\left(\theta, \frac{mg}{R}\right)$$

for some function f_2.

In both of these expressions for t, we have a time factor multiplied by a function of the two dimensionless quantities, θ and mg/R. The time factors are $(l/g)^{1/2}$ and $(ml/R)^{1/2}$. The first of these gives a time scale associated with the period of oscillation and the second gives a time scale associated with the damping due to air resistance. The quantity mg/R is a dimensionless combination of physical quantities and is an example of a *dimensionless group*. This is a useful concept in modelling physical systems. In fluid mechanics a very useful dimensionless group is the Reynolds number $(Re) = \rho ul/\mu$, where l is a characteristic length in the problem (e.g. the diameter of a pipe), ρ is the density of the fluid, μ is its viscosity and u is the fluid speed.

Example 4.4.1

The pressure p at a depth h below the surface of a fluid of density ρ is given by $p = \rho gh$, where g is the acceleration due to gravity. We can check that this equation is dimensionally correct as follows:

$$[p] = \left[\frac{\text{force}}{\text{area}}\right] = \frac{\mathrm{MLT}^{-2}}{\mathrm{L}^2} = \mathrm{ML}^{-1}\mathrm{T}^{-2},$$

$$[\rho] = \mathrm{ML}^{-3},$$

$$[g] = [\text{acceleration}] = \mathrm{LT}^{-2}$$

and

$$[h] = \mathrm{L}.$$

So $[\rho gh] = \mathrm{ML}^{-3}\mathrm{LT}^{-2}\mathrm{L} = \mathrm{ML}^{-1}\mathrm{T}^{-2}$ as required.

Example 4.4.2

The fluid viscosity μ is defined by the relation force per unit area $= \mu \times$ velocity gradient in the direction perpendicular to the area. If we take the dimensions of both sides of this relation, we have

$$\frac{\mathrm{MLT}^{-2}}{\mathrm{L}^2} = \frac{[\mu](\mathrm{LT}^{-1})}{\mathrm{L}}.$$

So

$$[\mu] = \mathrm{ML}^{-1}\mathrm{T}^{-1}$$

and μ is measured in kilograms per metre per second (kg m^{-1} s^{-1}). The dimensions of $(Re) = \rho ul/\mu$ are

$$(Re) = \frac{(ML^{-3})(LT^{-1})L}{ML^{-1}T^{-1}} = M^0L^0T^0,$$

i.e. (Re) is a dimensionless group as previously stated.

Example 4.4.3

When a particle falls under gravity in a viscous fluid, the drag that it experiences counteracts the acceleration due to gravity and after a while the particle's speed stops increasing. When this happens, we say that it has reached its 'terminal speed'. For a finite spherical particle, it is reasonable to suppose that the terminal speed v depends on the particle's diameter D, the viscosity μ of the fluid and the acceleration g due to gravity. Suppose also that v is directly proportional to the difference between the density ρ_1 of the particle and the density ρ_2 of the fluid. We assume that $v = kD^a(\rho_1 - \rho_2)\mu^b g^c$.

Taking dimensions of both sides, we have

$$\begin{aligned} LT^{-1} &= [k]L^a(ML^{-3})(ML^{-1}T^{-1})^b(LT^{-2})^c \\ &= [k]M^{1+b}L^{a-3-b+c}T^{-b-2c}. \end{aligned}$$

If k is a dimensionless number, then

$$1 + b = 0,$$

$$a - 3 - b + c = 1$$

and

$$-b - 2c = -1.$$

So b=–1, c=1, a=2 and our model is

$$v = \frac{k(\rho_1 - \rho_2)D^2 g}{\mu}.$$

Theoretical analysis using the laws of hydrodynamics leads to $v=(\rho_1-\rho_2)D^2g/18\mu$, an expression known as 'Stokes' drag'. This clearly fits with our expression obtained from simple dimensional considerations.

EXERCISES 4.4

1 Newton's gravitational law states that $F = Gm_1m_2/r^2$, where F is the magnitude of the gravitational force of attraction between two masses m_1

and m_2 separated by a distance r. What are the units of the gravitational constant G?

2 In a compressible fluid, the bulk modulus K is defined as

$$K = \frac{-(\text{change in pressure})/(\text{change in volume})}{\text{volume}}.$$

What are the dimensions of K?

3 Which of the following equations contain an error and which are dimensionally correct?

(a) $v^2 = u^2 + 2gz$.

(b) $p = \dfrac{\rho l u}{z}$.

(c) $F = -pA$.

(d) $p + \frac{1}{2}\rho u^2 = -mgz$.

p is a pressure, ρ is a density, u and v are velocities, F is a force, m is a mass, A is an area, and z and l are lengths.

4 The following models are meant to predict the volume flow rate Q of fluid through a small hole in the side of a large tank filled with fluid to a height H above the hole. Which one is dimensionally correct?

(a) $Q = \dfrac{AH^2}{g}$.

(b) $Q = A\sqrt{2gH}$.

(c) $Q = H\sqrt{2Ag}$.

A is the cross-sectional area of the hole and g is the acceleration due to gravity.

5 Check that the equation

$$\frac{\mathrm{d}p}{\mathrm{d}r} - \frac{\rho u}{r^2} = 0$$

is dimensionally correct, where p is pressure, ρ is density and u is velocity.

6 The velocity u of propagation of deep ocean waves is a function of the wavelength λ, acceleration g due to gravity and density ρ of the liquid. By assuming that $u = k\lambda^a g^b \rho^c$, where k is a dimensionless constant, find the relationship between u, λ, g and ρ.

4.5 SUMMARY

1 All variables apart from those which are pure numbers are measured in certain *units*. It is vital to keep note of these units

(a) when collecting data,
(b) when making your list of factors for your model and
(c) when testing your model against data.

The preferred system of units is the SI system but you may need to convert from and to non-SI units in any particular application.

2 Any physical quantity has a dimension which is a product of powers of the basic dimensions M, L and T. Check that all the equations in your model are dimensionally consistent.
3 Dimensional analysis can help in developing models. It shows how certain variables must be grouped together.
4 Dimensionless groups of variables are often useful in physical models. In particular, they give an indication of the relative importance of various physical influences.

5 MODELLING SKILLS

5.1 INTRODUCTION

It would not be possible to write down a list of all the skills which may be needed in developing mathematical models. They are many and varied: some of them may be described as intuitive, others come from long experience and practice, and some could be described as just plain common sense. In this chapter, we draw attention to some of the skills which are most commonly involved in modelling and which can be developed by practising with examples such as those presented here.

5.2 LISTING FACTORS

In chapter 3, we saw that the second step in the modelling process requires us to list all the factors that we can identify as being relevant to the problem. This may well be a long list and some of the factors will usually be more important than others. In our first run through the modelling cycle, we try to keep the model as simple as possible and this usually means reducing the list of factors to a manageable size. To do this, we go through the list and throw out the least relevant factors, retaining only the more important ones. We have to use our judgement plus whatever knowledge we have about the system that we are trying to model, including possibly any data that happen to be available. We may well find later that we have discarded some factors which ought to have been included. This will come to light at the validation stage when we find that the model gives an inadequate representation of reality. (Of course, this inadequency may be due to other reasons rather than to the omission of certain factors.)

On our next trip through the modelling cycle, we may well decide to include more factors, thereby complicating but hopefully also improving the model. Quite often the degree of importance of a particular factor cannot be judged at the outset but, when we have developed our model and used it to produce an answer, we can investigate what effect that particular factor has on the answer by varying the factor and noting what happens to the answer. This is also known as sensitivity analysis and we shall return to it again in chapter 6.

The word 'factor' is quite general and includes a number of different entities. A factor may be *quantifiable*, i.e. can be given a numerical value, or it may be a *quality* which can be named but not measured numerically. A factor could also be simply a *relationship* between two other factors. Quantifiable factors can normally be divided into variables, parameters and constants. *Constants* such as the speed of light have fixed values. *Parameters* have constant values for a particular problem but can change from problem to problem. For example, suppose that we are developing a model to solve a problem which involves a fluid. It is very likely that the density of the fluid would appear in our model and would thus be a parameter. We could use our model to predict what would happen if a fluid with a different density were used by changing the value of this parameter.

Variables can be discrete (i.e. capable of taking only certain isolated values such as integers) or continuous (i.e. capable of taking all values in a real interval). Variables can also be random or deterministic. It is often useful to distinguish between variables (or parameters) which are *inputs* to a model and those which are *outputs* from the model. In some cases, it will be possible, after working through the model, to express the output variables in terms of the inputs by mathematical expressions. In other cases the connection between outputs and inputs will have to be derived in the form of tables of numerical values. The answer required from the model will usually be the behaviour of all or some of the output variables or some value to be calculated from them.

In order to simplify the mathematics, it is normal to abbreviate the names of the variables and some thought should be given to choosing a suitable notation. It is a good idea to use symbols which remind us of the variables to which they refer, e.g. t for time, p for pressure and F for force. It is also sensible to conform to common usage and to denote angles by Greek letters such as α, β, γ, θ or ϕ and lengths by x, y, z or l. Keep symbols such as π and e for their usual mathematical meaning. When our model involves several similar variables, an indexed notation such as $X_1, X_2, X_3, \ldots$ or $A_{1,1}, A_{1,2}, \ldots$ may be appropriate. When a model is implemented on a computer, it is wise to lengthen the variable names—for example, use DIST for distance rather than just the single letter x. This helps to make the computer program understandable especially if there are many variables.

Having listed our factors and rejected the least important ones, our next step is to think about the relations between the factors, also bearing in mind the purpose of our model and its domain of validity. It can be useful at this stage to form *groups* of closely related factors, i.e. to list the factors under a number of headings. We can then try to formulate relationships between factors within a group before we try to relate the groups. The form of the relationship between any two or more factors will first emerge as a verbal statement which we must then translate into a mathematical statement. This step is difficult to carry out and we look at some of the skills involved in the remaining sections of this chapter.

Example 5.2.1

Problem statement

A water tank is supplied with water through a pipe from a distant reservoir. The reservoir receives its water from rainfall and inflow from rivers. Water is also lost from the reservoir through seepage into the soil and evaporation into the air. Water for domestic use is taken from the water tank through an outlet pipe.

Objective

A model is required which will predict the depth of water in the tank at any time, given all the relevant information (Fig. 5.1).

Model

We could list some of the relevant factors as follows.

Capacity of reservoir.
Depth of water in reservoir.
Rainfall rate.
River inflow rate.
Evaporation.

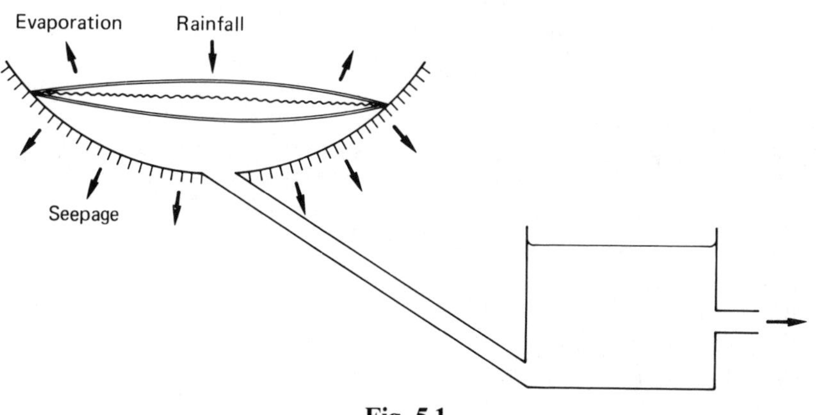

Fig. 5.1

Seepage.
Size of supply pipe from reservoir.
Flow rate through supply pipe.
Capacity of water tank.
Depth of water in the tank.
Size of outlet pipe from the tank.
Flow rate through tank outlet.

Some questions to ask at this stage are the following.

(a) Have we left out anything vital?
(b) Have we defined our factors meaningfully and precisely, e.g. what do we mean by the 'size' of a pipe?

In reply to question (a), we might add the difference in height between the reservoir and the tank as a factor which ought to have been included and, in reply to question (b), we could say that the pipe diameter is the relevant measurement although in the case of the supply pipe from the reservoir the pipe length might also be relevant. In the case of the outlet pipe from the tank, we also need to know the height of the outlet above the bottom of the tank.

One obvious way of grouping our factors would be as follows.

Factors concerning the reservoir.
Shape.
Capacity.
Water depth.
Rainfall rate.
River inflow rate.
Evaporation.
Seepage.

Factors concerning the connecting pipe.
Pipe diameter.
Pipe length.
Height difference between ends of pipe.
Flow rate in the pipe.

Factors concerning the tank.
Capacity.
Cross-sectional area.
Water depth.
Diameter of outlet pipe.
Height of outlet pipe above bottom of tank.
Flow rate through outlet.

If the purpose of the model is to predict the depth of water in the tank, then this is clearly going to be an *output* variable. What are the *input* variables? Before we try to answer this question, we can make things easier by firstly going through the factor list and identifying those factors which are either constants or parameters. The reservoir capacity, the dimensions of the connecting pipe, the height difference, the capacity of the water tank, the diameter of the outlet pipe and the height of the outlet pipe above the bottom of the pipe are all *parameters*. The remaining factors are all variables.

The next step in distinguishing input variables and output variables is to look for relationships within each group of factors. Can we see that some variables are direct consequences of other variables? Which variables are clearly independent of any other variables?

In the group of factors concerning the reservoir, the rainfall rate and the river inflow rate are obviously independent of the other variables. Are they independent of each other? Probably they are not because, if there has been a lot of heavy rainfall, the rivers will be swollen while in dry conditions the river inflow might dwindle to zero. Remember, however, that we should try to keep a model as simple as possible unless it proves to be inadequate. In our case, it will help us to assume that the rainfall rate and the river inflow rate are independent of each other. Their independence of any other variables means that they are *input* variables to the model.

The evaporation and seepage obviously have something to do with the amount of water in the reservoir at any moment although we could simplify matters by assuming them to be constants. More realistically, the evaporation rate depends on the area of water surface in the reservoir, which can be related to the depth of the water if we know the physical shape of the reservoir (we could model it as part of a sphere for example). Both the evaporation and (to a lesser extent) the seepage will also be affected by weather conditions, in particular the air temperature around the reservoir and the presence of wind. However, we have not included the weather in our model and we could justify this by saying that we have 'averaged out' the weather variations.

The depth of water in the reservoir is clearly a consequence of the flow of water into the reservoir and the flow through the connecting pipe into the water tank. The water depth is therefore an *output* variable from the model. If we decide to regard the seepage and evaporation as functions of the water depth in the reservoir, then these will also be *output* variables from the model.

The depth of water in the tank clearly depends on the flow rates in and out of the tank and on its dimensions and is therefore an *output* variable. The flow rate through the outlet is assumed to be controlled by the domestic users. We have to know their needs before we can calculate the flow rate through the outlet which is consequently an *input* variable for the model. We note that no outflow is possible if the water depth in the tank falls below the level of the outlet pipe.

To complete our list of factors, we give each a symbol and also indicate its units of measurement (Table 5.1).

Table 5.1

Description	Type	Symbol	Units
River inflow	Input	f_1	$m^3\ s^{-1}$
Rainfall	Input	f_2	$m^3\ s^{-1}$
Seepage	Output	s	$m^3\ s^{-1}$
Evaporation	Output	a	$m^3\ s^{-1}$
Reservoir depth	Output	d_1	m
Connecting pipe diameter	Parameter	D_1	m
Connecting pipe flow rate	Output	f_3	$m^3\ s^{-1}$
Height difference	Parameter	H	m
Connecting pipe length	Parameter	l	m
Tank area	Parameter	A	m^2
Tank capacity	Parameter	V	m^3
Tank water depth	Output	d_2	m
Outlet pipe diameter	Parameter	D_2	m
Height of outlet	Parameter	h	m
Outflow	Input	F	$m^3\ s^{-1}$

It may appear strange at first glance that the water outflow is classified as an input but remember that we are referring to *information* inputs and outputs and we are assuming that information on the use of the water is going to be available.

We are now in a position to formulate our model objectives in the standard 'given–find' form. It is simply this: *given* the inputs and the parameter values, *find* the outputs. In more detail, the objective of our model is to find the relationships between the outputs (seepage, evaporation, reservoir depth, tank water depth and rate of flow in connecting pipe) and the inputs (river inflow, rainfall and domestic water outflow) given the parameters (connecting pipe diameter and length, height difference, tank capacity, outlet pipe diameter and height of outlet). This completes the first step in the modelling cycle and we shall not continue further with this particular model.

EXERCISES 5.2

1 Make a list of all the factors which might be relevant in helping to decide whether or not to install double glazing in an existing house.

2 The headmaster of a new school is trying to decide how long the lessons should be. Write down all the factors that he may need to consider before making his decision.

3 What factors would you include in a mathematical model to help a shot putter to maximise the distance of his throw?

4 You are on the sixth floor of a tower block. How long do you expect to have to wait for a lift? What information do you need before you can answer this question? How many floors would you walk up or down rather than wait?

5.3 MAKING ASSUMPTIONS

To make progress with our model building, we must make some assumptions; in fact, keeping the 'building' metaphor in mind, if the factors are the building blocks, the assumptions provide the cement with which to put the structure together. The ultimate success or failure of our model will very likely depend on whether we make an apt choice of assumptions and this is where the expertise of an experienced modeller becomes evident. Having said how important it is to make suitable assumptions, we have to go on to admit that experience counts here, more than in any of the stages involved in modelling, and it is unfortunately difficult to give general advice to beginners.

A variety of assumptions may be necessary, e.g. the following.

1 Assumptions about whether or not to include certain factors.
2 Assumptions about the relative sizes of terms or the relative magnitudes of the effects of various factors.
3 Assumptions about the forms of relationships between variables.

Generally speaking, and especially when developing a new model for the first time, we try to choose assumptions which keep the model as *simple* as possible. Always take care to write down your assumptions clearly so that you are aware of them yourself and that later on when explaining your model to others they will be able to see exactly what assumptions have been made. Look back at the examples in chapter 2 and make a list of the assumptions which were made in each example. Do you think that they are reasonable?

Try to note all the *consequences* of your assumptions. Wherever possible, test your assumptions against data (see chapter 6). Beware of *implicit* assumptions which you may make without realising that you have made them. In the previous section, we did not assume anything about friction in the pipes. By leaving out pipe friction from the model, we did in fact assume that friction was negligible.

For models involving mechanics (there are some examples in chapter 8), we are helped by the fact that there are well tried and tested assumptions and we can follow the example of previous modellers from the time of Newton.

Closely allied with assumptions are modelling *decisions* which we usually have to make, the most important of which are as follows.

1 What is the appropriate level of detail to include in our model? We always have to struggle towards a compromise between the complexity of the real problem and the need to produce a limited but useful model within the finite time and resources available to us.
2 Should we try an analytic model or a simulation model?
3 Should we model the variables as discrete or continuous?
4 Should we use a stochastic or a deterministic model?

While debating these questions, remember the *purpose* of your model and do not try to be too ambitious.

5.4 TYPES OF BEHAVIOUR

When variables affect each other, the way that one variable behaves when another is varied is most conveniently expressed in terms of a graph with one variable plotted on the horizontal axis and the other on the vertical axis. We often think of the variable measured on the horizontal axis as the independent variable and the other as the dependent variable. In many problems the independent variable is time and we are looking at the time behaviour of our variables. Alternative ways of describing the relationship between two variables is by a mathematical formula or a table. In modelling, we often need to translate from one form to another. We can easily produce the graph corresponding to a particular formula but finding a mathematical formula corresponding to a graph or a table is not so easy and we look at ways of doing this in chapter 6.

In the case of input variables, we may have only a vague idea of how they behave, in which case we assume the simplest mathematical forms which give the right kind of behaviour as far as we can judge. Sometimes there are some data to guide us and, if a substantial number of data are available, we can use the methods of chapter 6 to fit an appropriate mathematical form.

It is particularly useful to consider the following questions. (We shall assume that t is the independent variable and y is the dependent variable.)

1 What happens at large t?
2 What happens at small t?
3 Are there any values of t for which y has a local maximum or minimum?
4 Are there any values of t for which $y=0$?
5 Are we interested in all values of t or only a certain range, say $t>0$ or $t_1<t<t_2$ for positive t_1, t_2?

In the case of output variables, whether we present their behaviour in graphical, formula or tabular form depends partly on personal preference and partly on how complicated our model is.

The following is a collection of simple types of behaviour which are very commonly used in modelling. More complicated types can be built up by combining these simple forms together. The letters a, b and ω represent parameters taking real positive values and we assume that t starts at 0 and takes real positive values only. Note that, apart from the linear case, the mathematical expressions quoted are just the simplest forms showing that particular behaviour. Many other mathematical expressions with a similar behaviour exist.

Linear (Fig. 5.2)

$$y = y_0 + at.$$

At $t = 0$, $y = y_0$.

Growing without limit (Fig. 5.3)

$$y = y_0 \exp(at).$$

y increases with t for all t. At $t = 0$, $y = y_0$.

Increasing to a limit (Fig. 5.4)

$$y = y_\infty(1 - \exp(-at)).$$

At $t = 0$, $y = 0$. At large t, y approaches y_∞ (because $\exp(-at)$ becomes negligible).

Decaying to a limit (Fig. 5.5)

$$y = y_0 \exp(-at) + b.$$

At $t = 0$, $y = y_0 + b$. y decreases for all t. At large t, y approaches b.

Simple maximum (Fig. 5.6)

$$y = at - bt^2.$$

$y = 0$ when $t = 0$ and when $t = a/b$. y has a local maximum at $t = a/2b$.

Maximum followed by tailing off (Fig. 5.7)

$$y = at \exp(-bt).$$

$y = 0$ when $t = 0$. y becomes 0 for large t. y has a local maximum at $t = 1/b$.

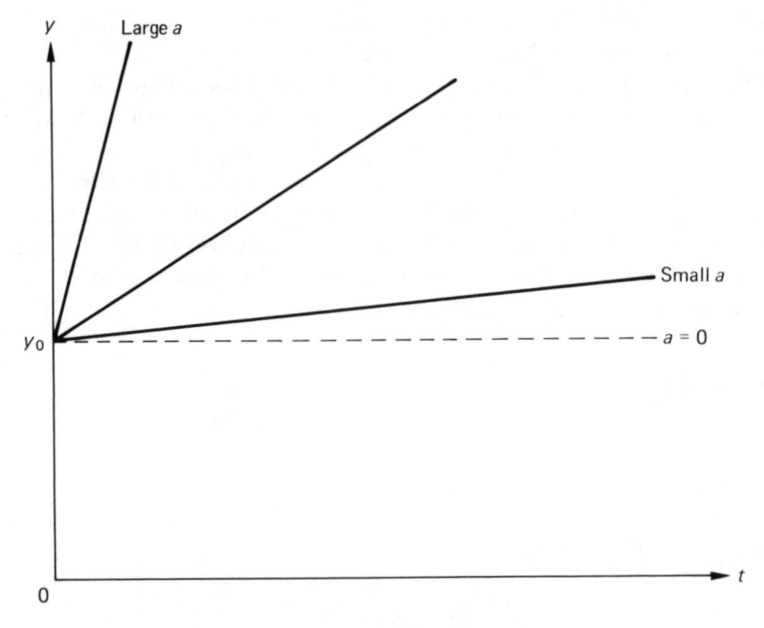

Fig. 5.2

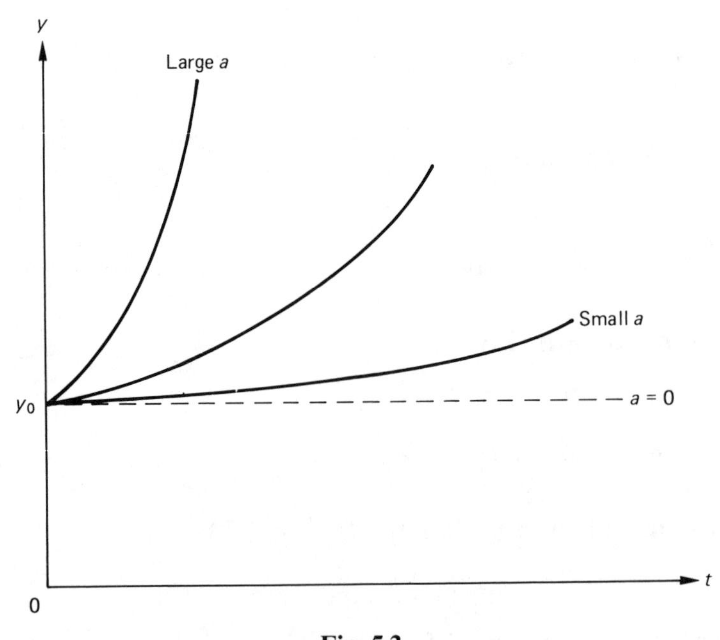

Fig. 5.3

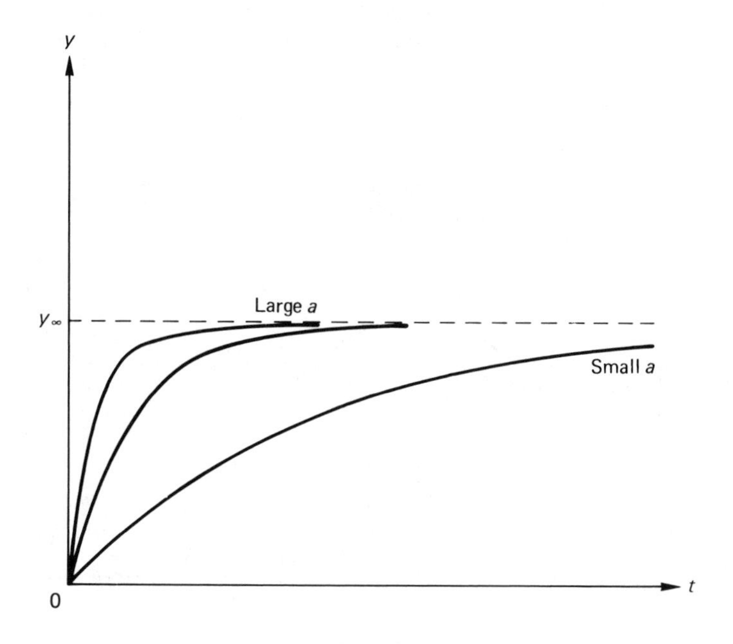

Fig. 5.4

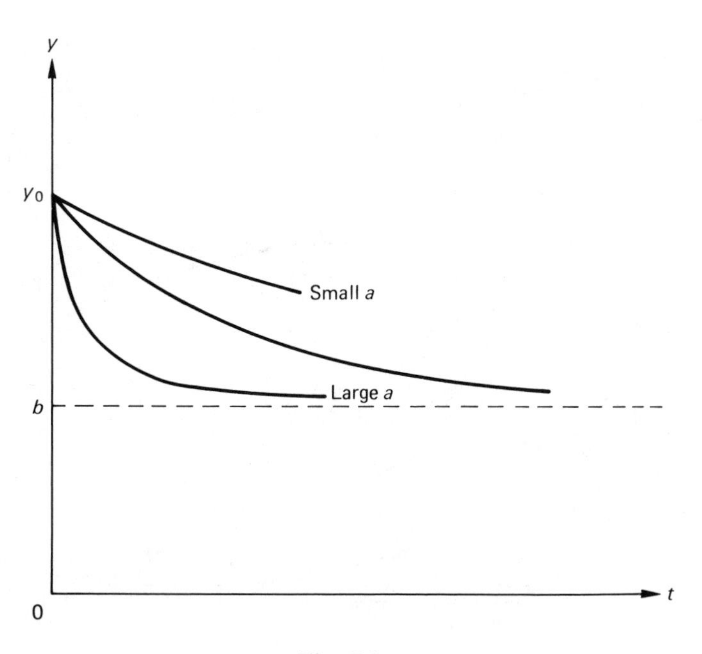

Fig. 5.5

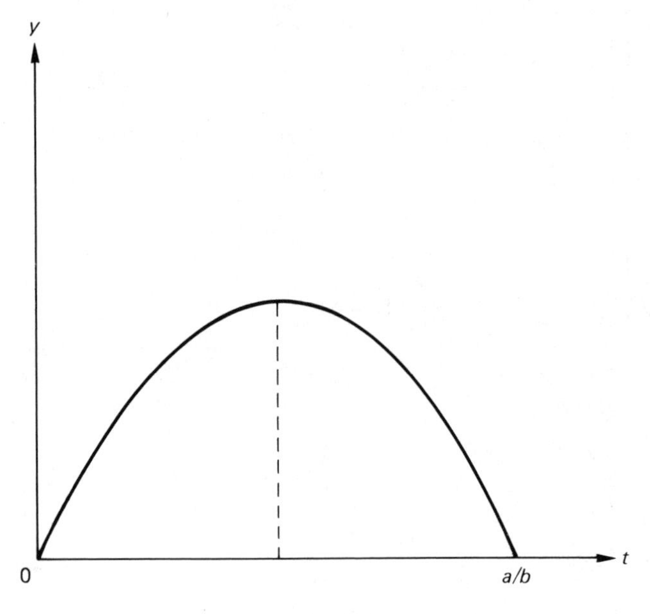

Fig. 5.6

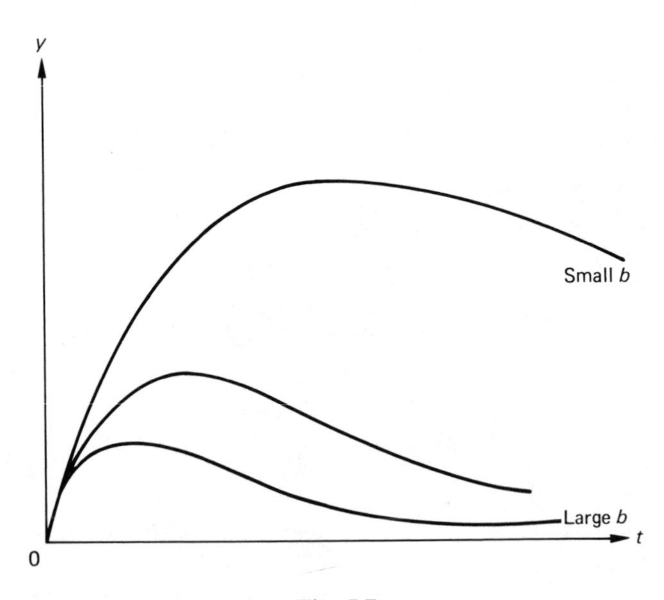

Fig. 5.7

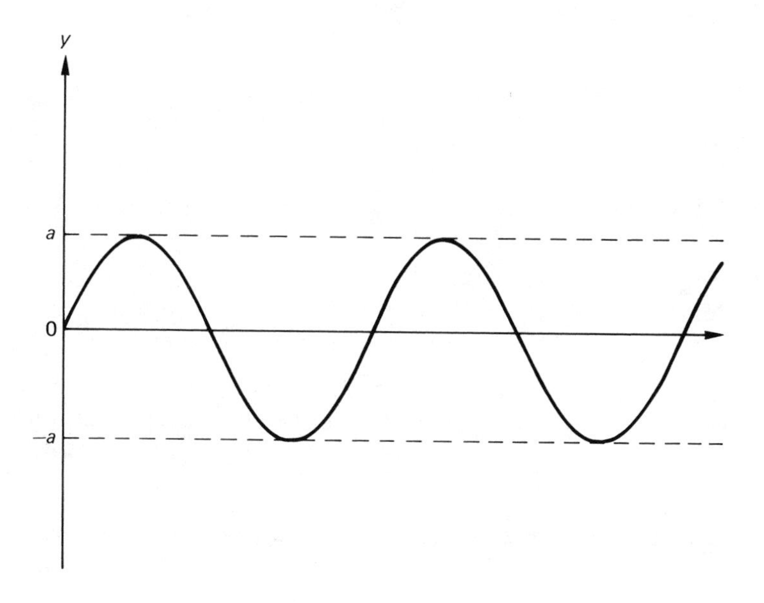

Fig. 5.8

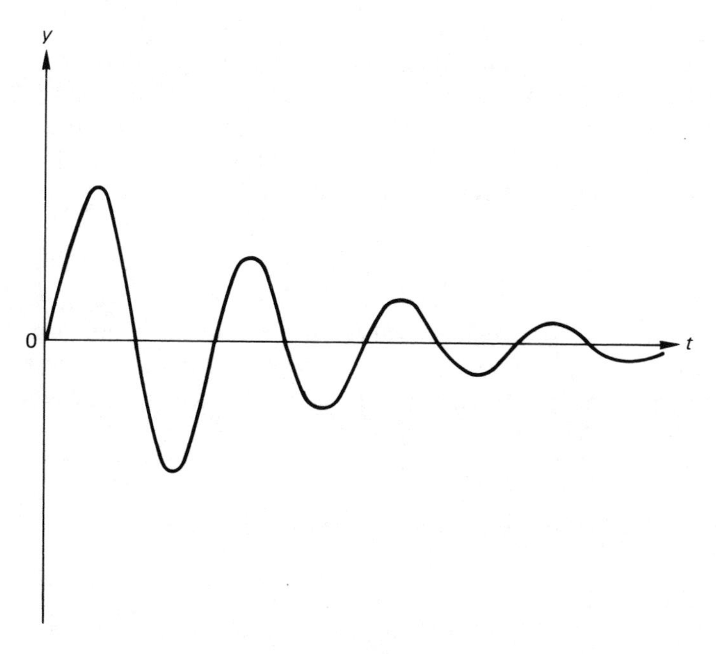

Fig. 5.9

Oscillatory (Fig. 5.8)

$$y = a \sin(\omega t).$$

$y = 0$ at $t = 0$ and at $t = n\pi/\omega$, where $n = 1, 2, 3, \ldots$ the period is $2\pi/\omega$ and the amplitude is a.

Decaying oscillations (Fig. 5.9)

$$y = a \exp(-bt) \sin(\omega t).$$

$y = 0$ at $t = 0$ and at $t = n\pi/\omega$ where the period is $2\pi/\omega$. The amplitude decreases with increasing t. The ratio of successive amplitudes is $\exp(-2b\pi/\omega)$.

Example 5.4.1

A culture of bacteria is growing rapidly. If its size now is 100 organisms and the population doubles in size every 5 min, what expression could we use for the population size at time t?

Solution

If we assume that $y = y_0 \exp(at)$, then at $t = 0$ we have $y = y_0 = 100$. At $t = 5$, $y = 200 = 100 \exp(5a)$. Therefore, $a = \frac{1}{5} \ln 2 \approx 0.139$. So a possible model is $y = 100 \exp(0.139t)$. Note that, as it stands, this model implies that the population continues to grow without limit. In practice, limitations of space and/or food supply will prevent this and we need to restrict the domain of validity of our model to some finite interval of time.

Example 5.4.2

Suppose that we wish to model the daily average number of hours of sunshine at a particular location. If we start measuring from the winter minimum when the average number of hours of sunshine is $y_{\min}$ and let t be the time in days from this point, then a suitable simple model might be

$$y = y_{\min} + b \sin^2(\omega t).$$

What values should we take for b and ω?

Solution

The period of $\sin^2(\omega t)$ $(= \frac{1}{2}[1-\cos 2\omega t])$ is $\pi/\omega = 365$ days; so $\omega = \pi/365 \approx 0.0086$. If $y_{\max}$ is the value at the summer peak, this occurs when $\omega t = \pi/2$, i.e. when

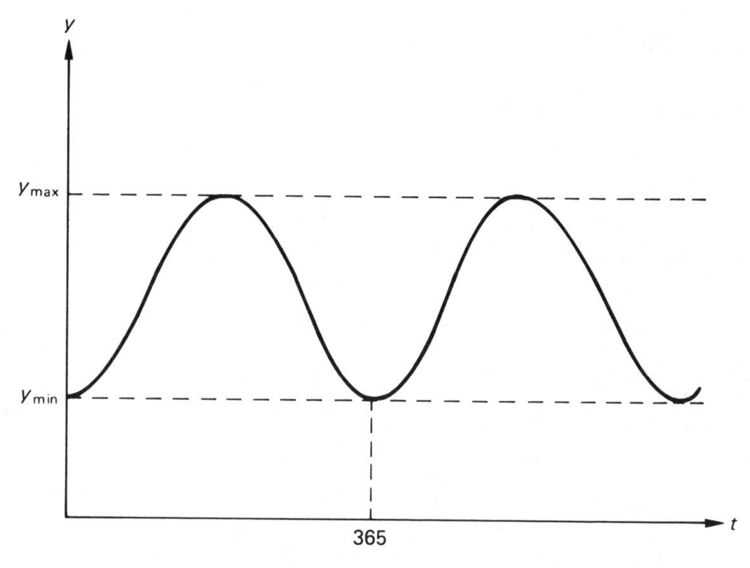

Fig. 5.10

$y = y_{max} = y_{min} + b$. So $b = y_{max} - y_{min}$ and the model is (Fig. 5.10)

$$y = y_{min} + (y_{max} - y_{min}) \sin^2(0.0086t).$$

Example 5.4.3

In many situations, if the price of an article is increased, the number of articles bought decreases. A simple model for this kind of behaviour is $y = y_0 - ap$, where y is the number of articles sold when the price is p. What values should we take for y_0 and a?

Solution

When the price is increased by one unit, a is the decrease in y. y_0 is the number theoretically sold when $p = 0$, i.e. when the articles are given away free! This is somewhat artificial and a better idea would be to take two points, i.e. to estimate the numbers that would be sold at two different prices p_1 and p_2. Suppose that we estimate these to be y_1 and y_2. Then $y_1 = y_0 - ap_1$ and $y_2 = y_0 - ap_2$; so $y_2 = (y_1 - y_2)/a(p_1 - p_2)$ and the model is

$$y = \frac{y_1 - y_2}{a(p_1 - p_2)} - ap.$$

An appropriate domain of validity for the model would be the interval $p_1 < p < p_2$. The model *could* be used for p values outside this range but only at the user's risk.

EXERCISES 5.4

1 Which of the following expressions

(i) are increasing for all x,
(ii) are very large for small x,
(iii) become very small at very large x and
(iv) are unbounded for large x?

(a) $x + \frac{1}{x}$.
(b) $1 - \exp(-x)$.
(c) $x \exp(-x)$.
(d) $\frac{1}{1+x^2}$.
(e) $\frac{1}{1+\exp x}$.
(f) $\frac{1}{1+\exp(-x)}$.
(g) $\frac{\exp x}{x^2}$.
(h) $\frac{1+\exp(-x)}{1-\exp(-x)}$.

2 Do the following expressions increase or decrease

(i) as a increases,
(ii) as b increases and
(iii) as c increases?

(a) $\frac{a}{1+b/c}$.
(b) $\frac{b}{c} - a + \frac{b}{a}$.
(c) $\frac{abc}{a+bc}$.

3 Discuss the effects of changing the parameters a, b and c in the following expressions.

(i) $a + b\exp(-cx)$.
(ii) $c + ax\exp(-bx)$.

5.5 TRANSLATING INTO MATHEMATICS

Care must be taken when selecting an appropriate mathematical form corresponding to a verbal statement concerning variables. For example, if one variable y is stated to be directly proportional to another variable x (sometimes symbolised as $y \propto x$), the appropriate mathematical expression to represent the relationship is $y = kx$, where k is the constant of proportionality. The appropriate value of k can be derived if some sample data of x and y values are available.

If y is proportional to x_1 and proportional to x_2, then the appropriate form could be $y = kx_1x_2$. Note that this implies that y is doubled whenever x_1 or x_2 is doubled for example, which would not be the case for an expression such as $y = k_1x_1 + k_2x_2$. The second form would be appropriate for a situation such as 'y is increased by an amount k_1 for every unit increase in x_1 and by an amount k_2 for every unit increase in x_2'. We came across a situation such as this in Example 2.2.6 in chapter 2.

The statement 'y decreases as x increases' could be interpreted as a linear relationship $y = y_0 - ax$ $(a > 0)$ or as inverse proportion $y = k/x$. We would need either to obtain more information about the actual relationship between x and y or to make our own assumptions about which form is more appropriate or whether some other form should be used.

Example 5.5.1

Suppose that an ice cream seller at a summer fair guesses that the amount A of ice cream that he will sell is going to be

(a) proportional to the number n of people who come to the fair,
(b) proportional to the temperature excess over 15 °C and
(c) inversely proportional to the selling price p.

What would be an appropriate model for A?

Solution

These assumptions would lead to a model of the form $A = kn(T - 15)/p$, where T is the actual temperature in degrees Celsius. Note that the model is valid for $T \geqslant 15$.

EXERCISES 5.5

1 A variable w is related to two other variables x and y in such a way that w is proportional to x and also proportional to y. Which of the following correctly expresses the relationship?

(a) $w = a(x + y)$, a constant.
(b) $w = ax + by$, a and b constants.
(c) $w = axy$, a constant.

2 A variable y depends on two other variables w and z. The following facts are known.

(i) When w increases, y decreases.
(ii) When z increases, y also increases.
(iii) When w and z are both zero, y is also zero.

Which of the following models are consistent with facts (i), (ii) and (iii)?

(a) $y = aw + bz$, a and b constants > 0.
(b) $y = bz - aw + c$, a, b and c constants > 0.
(c) $y = \frac{cz}{w}$, c constant > 0.
(d) $y = cwz$, c constant > 0.
(e) $y = az - bw$, a and b constants > 0.

3 When a fluid flows through a pipe, the frictional force F between the pipe wall and the fluid is assumed to be proportional to the length L of the pipe and the square of the fluid speed U. It is also assumed to be inversely proportional to the diameter D of the pipe. Write down an expression for F in terms of L, U and D and involving a constant k. What are the dimensions of k. In what units would k be measured?

5.6 CHOOSING MATHEMATICAL FUNCTIONS

Consider the following common situations from everyday life.

Case 1 It rains suddenly for 20 min, the rain increasing in intensity before stopping.
Case 2 You are selling ice cream on a hot day; demand is at its greatest when the temperature is highest.
Case 3 You need helpers for the youth club jumble sale on Saturday morning; sales pressure is at its greatest immediately you open for business.

In all three cases, we can see that the effects represented are probably not constant throughout the events. If these phenomena occur as constituent parts of larger mathematical models, then we shall have to speculate on the behaviour of each before inserting the mathematical representation into our model. For example, the selling of ice cream could be investigated from the profit and loss point of view over a long period. Rainfall rates will contribute to models on reservoir collection, drainage problems and so on.

The common feature of the three situations above is that within each there is a *rate of change* in some quantity which is not constant over a period of time. Also in each case, the accumulated totals of the quantities are probably known, i.e. in Case 1 we can measure the total amount of rain collected in 20 min, in Case 2 we have a rough idea of how much ice cream can be sold per day, and in Case 3 we can count, or estimate from past experience, how many people will come to the jumble sale. Therefore, if the 'quantity' is denoted by Q, and time is denoted by t, then we have, from calculus, the relation

$$\int_{\text{start}}^{\text{end}} \mathrm{d}Q(t)\,\mathrm{d}t = Q(\text{end}) - Q(\text{start}).$$

From the modelling point of view, it is the form $\mathrm{d}Q/\mathrm{d}t = Q'(t)$ that is of most interest. In section 5.4, we have shown how the behaviour of certain mathematical functions can be represented. Now we shall build on that work by looking at various possibilities for representing the three situations outlined above.

Case 1

Suppose that, after 20 min of heavy rain, $\frac{1}{2}$ in of rain has fallen. You immediately think of rain falling over an *area*, but weather centres prefer to use the height measurement only. Our objective is to model the rate of rainfall by a suitable function $R'(t)$, and there are of course many possible choices.

(a) Steady continuous rain at a constant rate. This is shown in Fig. 5.11.
(b) It could start to rain slowly before picking up to a maximum and then subsiding again, over the 20 min period, with the total amount collected remaining at $\frac{1}{2}$ in. If a steady linear increase in the rate is taken followed by a similar decrease, then we arrive at Fig. 5.12. Now the maximum rainfall rate is 0.05 in min^{-1} so that we still get a total of $\frac{1}{2}$ in collected (check the area under the graph).
(c) There is no need to stop at the situation modelled by (b), particularly if the rain is more or less steady apart from the start and finish. Suppose that the rate increases steadily for 2 min and then remains constant for the next 16 min before decreasing again until it stops after 20 min. This is represented in Fig. 5.13. The maximum rainfall rate would then be approximately 0.028 in min^{-1}. (How did we get that figure?)

Now, as a constituent part of some larger model, perhaps resulting in a differential equation, it is the functional form of $R'(t)$ that is needed.

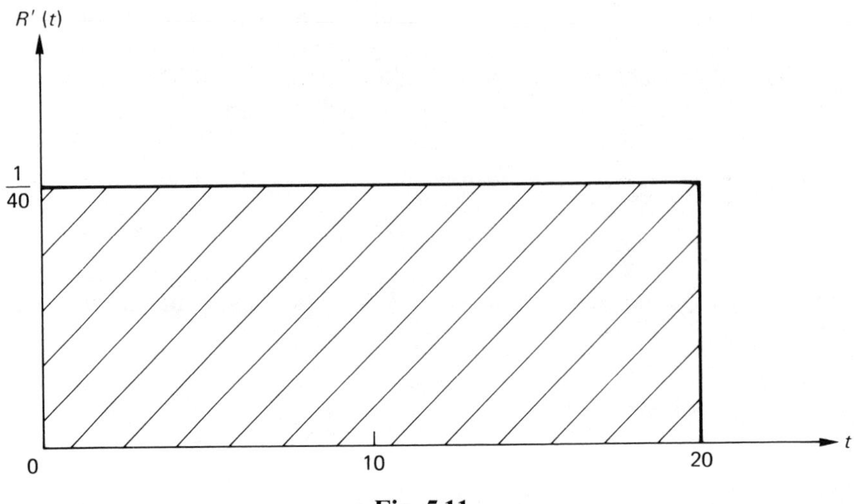

Fig. 5.11

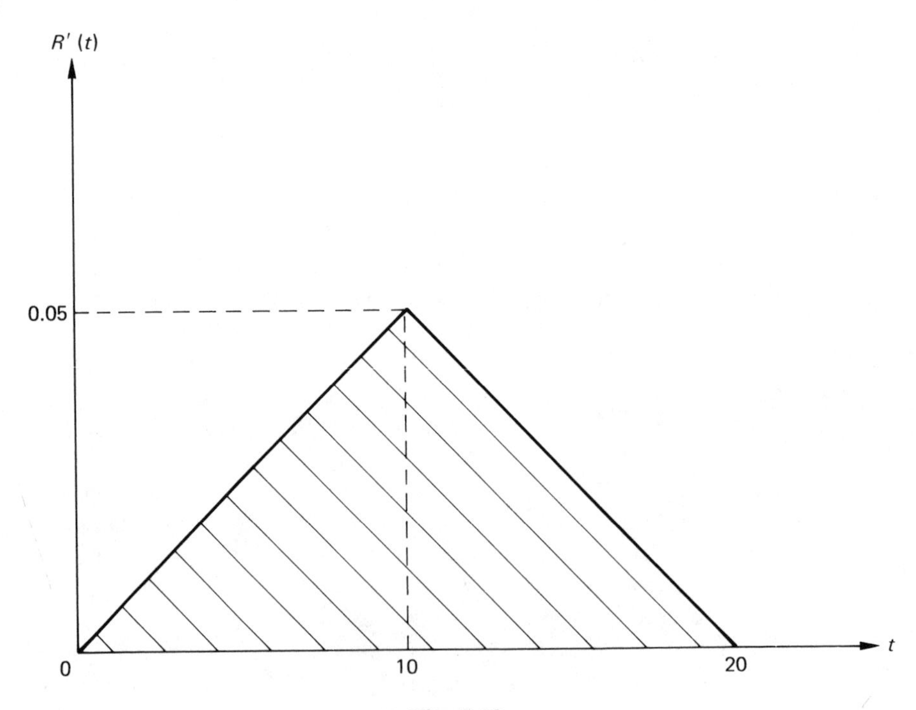

Fig. 5.12

For (a), we have

$$R'(t) = \frac{1}{40}, \qquad 0 < t < 20.$$

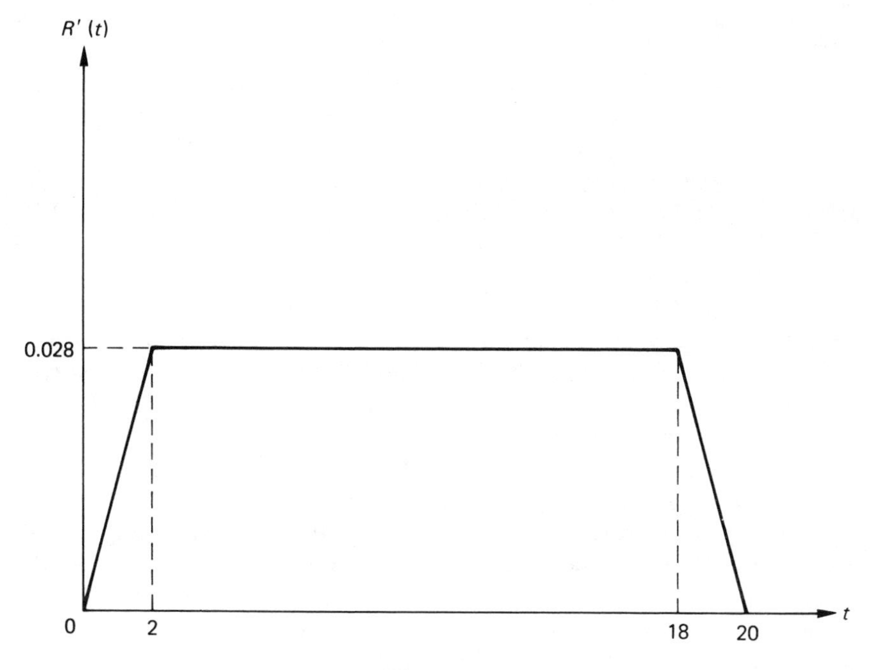

Fig. 5.13

For (b) we have the functional form given by two separate linear forms, each holding over different ranges of time t:

$$R'(t) = \begin{cases} 0.005t, & 0 < t < 10, \\ 0.1 - 0.005t, & 10 < t < 20. \end{cases}$$

Note that the two formulae agree at $t = 10$ (otherwise we would have a discontinuous function).

For (c) we have the functional form given by three separate parts, each holding over different ranges of t values:

$$R'(t) = \begin{cases} 0.013\,89t, & 0 < t < 2, \\ 0.027\,78, & 2 < t < 18, \\ 0.278 - 0.013\,89t, & 18 < t < 20. \end{cases}$$

(Note that we may not need all the decimal places.)

Again, check that these forms match up at the change-over points $t = 2$ and $t = 18$; so there are no breaks. It is quite common in modelling to find that a function is best represented by a formula made up from different forms in different parts of the range as in this example. Do not think that one single formula covering the whole range has to be found (although this is very convenient if it can be done).

Case 2

For the ice cream sales, much the same set of mathematical functions *could* be used. However, it is more likely here that sales will build up to a peak in the middle of the day and then subside again, in a 'smoother' way. Note that we are modelling the amount of ice cream sold as a continuous variable although it is sold in discrete lumps and is therefore a discrete variable. We assume that the approximation is sufficiently accurate for the purpose of the model. The amount of rain collected at any time is of course very accurately represented as a continuous variable.

Suppose data are given that 1000 ice cream cones are sold on a hot day. The kiosk is open for 8 h continuously from 10.00 am until 6.00 pm. Let $I(t)$ represent the number of ice cream cones sold up to time t measured in hours with $t = 10$ corresponding to 10.00 am. To model the gradual rise in sales, we are interested first in selecting a suitable function for the rate $I'(t)$ of sales. One possible selection is based on the use of the sine function.

From our knowledge of the behaviour of the sine function given in section 5.4, we select $I'(t) = a\sin(\omega t)$ which has a period $2\pi/\omega$ and amplitude a. A moment's thought tells us that this is inadequate since $\sin(\omega t)$ will take up negative values for certain t values. A better choice will be provided by $I'(t) = a\sin^2(\omega t)$, but t has to be scaled to run from 10:00 to 18:00 h and we want I' and I to be zero at each end value of t. The required form of behaviour is shown in Fig. 5.14.

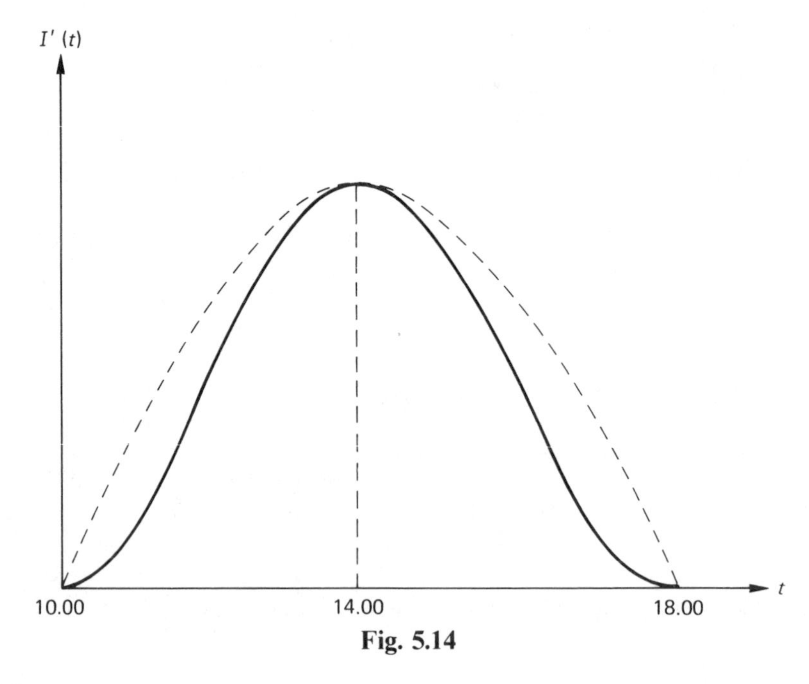

Fig. 5.14

Note that we do not choose a quadratic curve (shown as a broken curve in the figure) because it has steep slopes at the end points. After some thought, we decide on

$$I'(t) = \begin{cases} a \sin^2\left(\dfrac{\pi(t-10)}{8}\right), & 10 < t < 18, \\ 0, & \text{otherwise.} \end{cases}$$

(Check that this satisfies the required conditions.)

We now need to calculate the parameter a so that the total sales over the whole day amount to 1000 cones. This is easily done by integration

$$\int_{10}^{18} a \sin^2\left(\frac{\pi(t-10)}{8}\right) \mathrm{d}t = 1000,$$

i.e.

$$\frac{a}{2}\int_{10}^{18} \left\{1 - \cos\left(\frac{2\pi(t-10)}{8}\right)\right\} \mathrm{d}t = \frac{a}{2}\left[t - \frac{4}{\pi}\sin\left(\frac{2\pi(t-10)}{8}\right)\right]_{10}^{18} = 4a = 1000.$$

So finally our model is

$$I'(t) = 250 \sin^2\left(\frac{\pi(t-10)}{8}\right).$$

This means that at the height of the day at 2.00 pm we are selling at the rate of 250 cones every hour, i.e. four cones every minute.

Case 3

The jumble sale problem is somewhat different. This time we are given that 500 customers are expected; many of them will be waiting for the doors to open at the start. Suppose that the sale runs from 10.00 am until 12.00 noon. We shall need most helpers at the start to deal with the rush. Later, as many of the most sought-after items have been sold, demand will slacken off. Suppose that we are interested in the rate of entry of customers. Denote this by $Q'(t)$ where t is the time in hours. Then again, as we know the total to be 500, there is the relation

$$\int Q'(t)\,\mathrm{d}t = 500. \tag{5.1}$$

In Fig. 5.15, we speculate on the inflow. We have indicated that $Q'(t)$ is very high at the start, remains constant for the rest of the first hour and then decreases steadily to zero at the end.

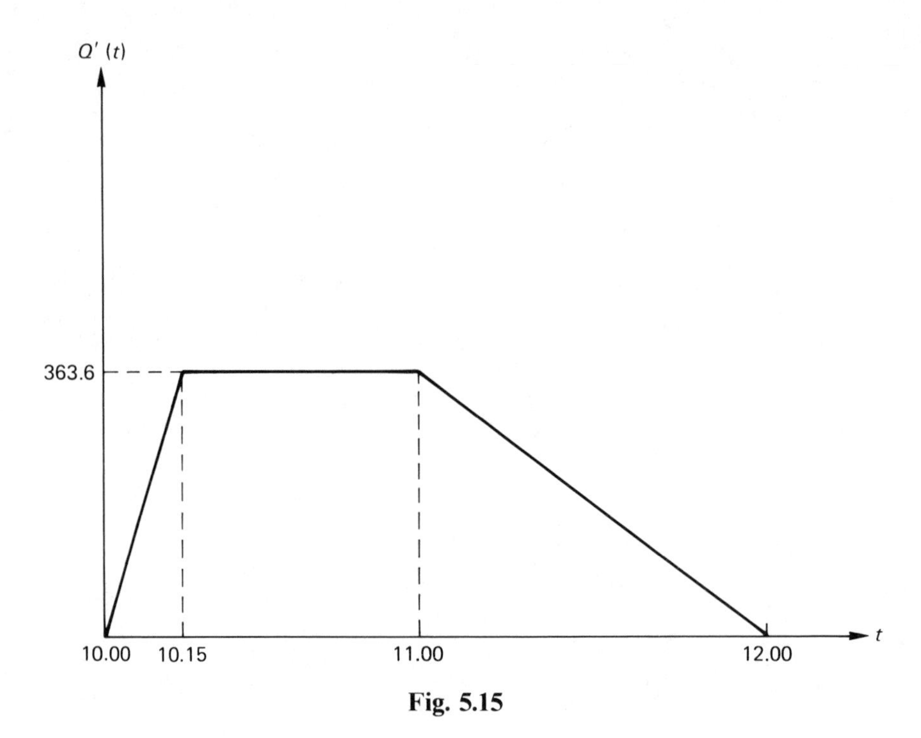

Fig. 5.15

As before, $Q'(t)$ requires more than one equation for its specification:

$$Q'(t) = \begin{cases} 363.6(t-10), & 10.00 < t < 10.15, \\ 363.6, & 10.15 < t < 11.00, \\ 363.6(12-t), & 11.00 < t < 12.00. \end{cases}$$

The factor 363.6 is calculated from equation (5.1) above. Modelling is an activity where we want to represent trends and relations between quantities, and usually at this stage we do not want a parameter quoted to several decimal places. This means that the slightly clumsy value of 363.6 might well be replaced by 360.0, even though the customer total would then be not quite equal to 500. We do not have to use the form of $Q'(t)$ just derived. You may wish to model the form in Fig. 5.16.

This form suggests a mathematical representation $at \exp(-bt)$. There are now two parameters to fix. We have the tailing off that we want, but we have to realise that Q' will never actually reach zero as t increases in this model. This alternative is left for you to finish, not forgetting that the time range is $10.00 < t < 12.00$ and also that the integral relation holds:

$$500 = \int_{10}^{12} a(t-10)\exp[-b(t-10)]\,\mathrm{d}t.$$

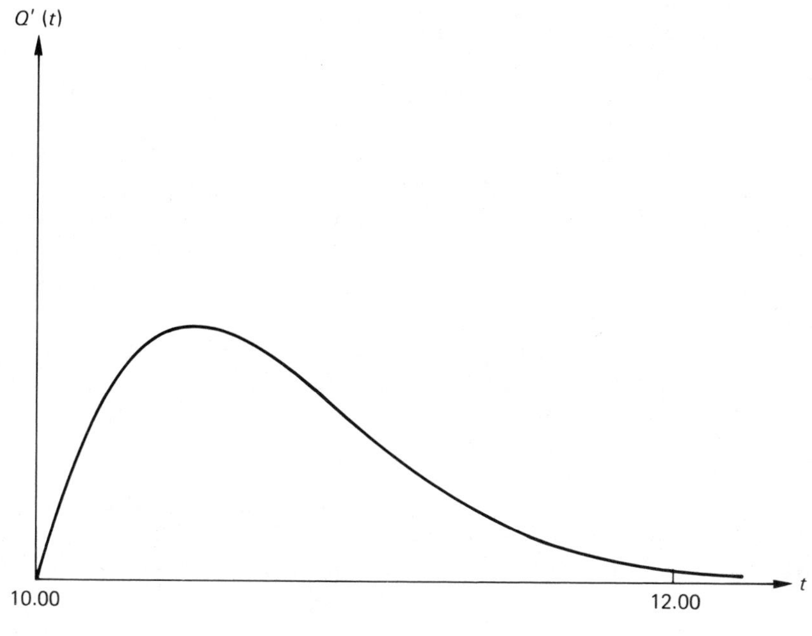

Fig. 5.16

EXERCISES 5.6

1 The air temperature just above the ground at a particular point on the Earth often varies in a periodic manner over a 24 h cycle. The daily mean value also varies with the seasons, i.e. over an annual cycle. If the time t is measured in hours what would be an appropriate mathematical model for the temperature as a function of t?

2 In Example 2.2.6 ('price war'), we assumed that the sales of both petrol stations would increase if they dropped their prices, or in other words the total petrol sales in their area would increase. Suppose instead that the local sales volume is constant so that the two garages are competing for larger shares of the same market. Suppose also that each garage's daily sales figure is inversely proportional to the price at which they sell. Obtain mathematical expressions for each garage's daily sales in terms of their selling prices x and y.

3 For a certain type of tree the rate of growth is slow in winter and greatest during the summer months. The rate of growth also lessens every year from a maximum in the first year until there is virtually no growth at all after 10 years. What would be a suitable mathematical model for the rate of growth?

5.7 RELATIVE SIZES OF TERMS

A model will often combine together a large number of variables, some with more effect than others on the final result. There is no point in including in the model more variables than are necessary to give an answer with the required amount of accuracy. If a particular variable makes an insignificant contribution to the answer when compared with the contributions of other variables, then it makes sense to abandon that variable especially if this simplifies the model. The same principles apply to expressions and equations which appear in the model. If an expression or equation involves a number of *terms*, it is often instructive to calculate roughly the relative sizes of the terms. Dropping the least significant terms will often simplify the mathematics considerably with no serious effect on the accuracy of the model.

Suppose that a variable x is known to have a value around 10. We use the notation $x \sim O(10)$ to mean 'x is of the order of 10' or 'x is in tens rather than in hundreds'. We have $1/x = x^{-1} \sim O(10^{-1})$ and $x^2 \sim O(10^2)$ so that in the expression $y = 1/x + x^2$ we have a sum of two terms whose magnitudes are $O(10^{-1})$ and $O(10^2)$. The second term is three orders of magnitude (a factor of 10^3) larger than the first; so, if we replace the original expression by $y = x^2$, the relative error that we are making (see section 6.4) should be about 10^{-3}. For example, suppose that $x = 8$; then $1/x + x^2 = 64.125$, while $x^2 = 64$ and the relative error is $0.125/64 \approx 0.002$. If the original expression had been

$$y = \sqrt{\frac{1}{x} + x^2} \exp\left(\frac{1}{x} + x^2\right),$$

we could replace it by the considerably simpler version

$$y = x \exp(x^2)$$

without much loss of accuracy.

In experiments containing several variables and parameters, we must evaluate the order of magnitude of each term. Generally the approximation that we get depends on which variable is smallest and it is sometimes useful to indicate this using a notation such as (x small). For example, if

$$z = 1 + xy^2 + x^2,$$

then we can say that

$$z = 1 + xy^2 \text{ (}x\text{ small)}$$

and

$$z = 1 + x^2 \text{ (}y\text{ small)}.$$

In each case we have dropped squares (and higher powers if there had been any) of the variables concerned. Consequently what we have obtained are referred to as 'first-order approximations'. Second-order approximations are obtained by dropping powers higher than second. In the above example, if both x and y are small, then

$$z = 1 \ (x \text{ and } y \text{ small})$$

is the first-order approximation and

$$z = 1 + x^2 \ (x \text{ and } y \text{ small})$$

is the second-order approximation for z.

Series expansions such as Maclaurin series are useful for obtaining low-order approximations. For example, if

$$y = \frac{1}{1+x} = (1+x)^{-1} = 1 - x + x^2 - x^3 + \ldots,$$

we can write

$$y = 1 - x$$

to first order in x. Similarly, if

$$y = \cos x = 1 - \frac{x^2}{2} + \frac{x^4}{24} + \ldots,$$

then

$$y = 1 - \frac{x^2}{2} \ (x \text{ small}).$$

This is a second-order aproximation.

Example 5.7.1

If

$$y = \frac{\exp(-x)}{1-x}$$

we can write

$$\begin{aligned} y &= (1-x)^{-1} \exp(-x) \\ &= (1 + x + x^2 + \ldots)\left(1 - x + \frac{x^2}{2} + \ldots\right) \\ &\approx 1 + \frac{x^2}{2} \ (x \text{ small}). \end{aligned}$$

Example 5.7.2

If

$$y = 1 + a\sin(\omega x) + a^2\omega^2\cos(\omega x),$$

then

$$y = 1 + a(\omega x - \tfrac{1}{6}\omega^3x^3 + \ldots) + a^2\omega^2(1 - \tfrac{1}{2}\omega^2x^2 + \ldots).$$

So

$$y = 1 + a^2\omega^2 + a\omega x \ (x \text{ small}),$$

$$y = 1 + a\sin(\omega x) \ (a \text{ small})$$

and

$$y = 1 + a\omega x \ (\omega \text{ small}).$$

EXERCISES 5.7

In the following questions the parameters a, b and c have the values $a = 0.1$, $b = 0.01$, $c = 0.001$ and the variable x can be assumed to lie in the interval $[1, 10]$.

1 Identify the smallest and the largest terms in each of the following expressions.

(i) $\dfrac{x}{b} + \dfrac{a}{x} + c.$

(ii) $\dfrac{ax}{b} + bx^2 + ab.$

(iii) $\dfrac{a}{x} + \dfrac{bx^3}{a} + \dfrac{c}{ax^2}.$

2 Simplify the following expressions by rejecting the smallest term in each.

(i) $x^2 - 2cx + a.$

(ii) $ac + bx^2 + \dfrac{a}{x}.$

(iii) $b\sin x + a\sqrt{x}.$

3 Simplify the following by rejecting all terms of order 10^{-n} with $n > 2$.

(i) $bx^3(1 - b^3x + cx - x^2c^2)$.

(ii) $\dfrac{b^3 + 2c/x + a^2}{c^3x + b}$.

(iii) $\sqrt{cx} + c \sin x + b \exp(-x)$.

4 Simplify the following as far as possible by retaining only the largest terms.

(i) $\dfrac{x^2 + ax}{b + cx + x^2/c}$.

(ii) $\dfrac{(ax + c \sin^2 x)(b^3x + a)}{bx + ac}$.

(iii) $\dfrac{a}{x^2} + \dfrac{10b}{x} + cx$.

5.8 REDUCING THE NUMBER OF PARAMETERS

Suppose that our model requires us to find x from the equation

$$ax^7 + bx = c,$$

where a, b and c are parameters. We cannot do this algebraically but we could substitute particular numerical values for a, b and c and solve the resulting non-linear equation by some numerical technique such as the Newton–Raphson method. We could arrange for a computer program to carry out the solution for a range of values of a, b and c. This would produce several pages of output. A neater method is first to reduce the number of parameters as follows. Let $x = \alpha u$ and $c = \beta v$; then the equation reads

$$a\alpha^7u^7 + b\alpha u = \beta v.$$

Divide by $a\alpha^7$ to give

$$u^7 + \left(\frac{b}{a\alpha^6}\right)u = \left(\frac{\beta}{a\alpha^7}\right)v.$$

Let the bracketed terms be equal to 1, i.e.

$$\frac{b}{a\alpha^6} = 1 \qquad \text{and} \qquad \frac{\beta}{a\alpha^7} = 1$$

or

$$\alpha = \left(\frac{b}{a}\right)^{1/6} \qquad \text{and} \qquad \beta = a\left(\frac{b}{a}\right)^{7/6} = \left(\frac{b^7}{a}\right)^{1/6}.$$

The equation that we have to solve is now

$$u^7 + u = v.$$

We can write a computer program to find u corresponding to a range of values of v, thus producing one page of output rather than several dozens. For particular values of the original parameters a, b and c, we find the solution x by calculating $v = c/\beta = c(a/b^7)^{1/6}$, looking up our u–v table for the solution u, and finally calculating

$$x = \alpha u = \left(\frac{b}{a}\right)^{1/6} u.$$

Graphical output is neater too, requiring only one u–v graph instead of pages of graphs for various a, b and c values.

Note that what we have done in this example is to 'de-dimensionalise' the equation. The bracketed terms $(b/a\alpha^6)$ and $(\beta/a\alpha^7)$ are examples of dimensionless groups. We cannot set more than two such brackets equal to 1; so any further brackets will remain in our equations as parameters. However, a reduction from three parameters to one is still very useful. This idea is especially useful with differential equations.

Example 5.8.1

Suppose that x is to be found from the equation

$$A(ax + b)^{1/3} + kx = c,$$

where all the other letters represent parameters. First let $ax + b = u$ so that the equation becomes

$$Au^{1/3} + \frac{k}{a}(u - b) = c$$

or

$$Au^{1/3} + \frac{ku}{a} = d$$

where $d = c + kb/a$. Let $u = \alpha v$ and $d = \beta w$; then

$$A\alpha^{1/3}v^{1/3} + \frac{k\alpha v}{a} = \beta w$$

or

$$v^{1/3} + \left(\frac{k\alpha^{2/3}}{Aa}\right)v = \left(\frac{\beta}{A\alpha^{1/3}}\right)w.$$

Let the bracketed terms equal 1, i.e. $k\alpha^{2/3}/Aa = 1$ and $\beta/A\alpha^{1/3} = 1$. The equation to be solved is $v^{1/3} + v = w$, i.e. an equation containing only one parameter, w. One graph of v against w contains all the information that we need. Returning to the original equation, if we wish to find x for a particular set of values of A, a, b, k and c, we calculate w from

$$w = \frac{d}{\beta} = \left(c + \frac{kb}{a}\right)\left(\frac{k}{A^3 a}\right)^{1/2}$$

and then read the corresponding value of v from the graph. The solution for x is

$$x = \frac{u - b}{a} = \frac{1}{a}\left[\left(\frac{Aa}{k}\right)^{3/2} v - b\right].$$

Example 5.8.2

Many mathematical models of physical systems involve periodic oscillation and a particularly common model is the elementary spring–mass–damper system. This is composed of a mass m attached to a spring and a damper. The spring exerts a restoring force proportional to its extension while the damper exerts a resistive force proportional to the speed of the mass. Application of Newton's second law of motion leads to the following differential equation for the position $x(t)$ of the mass at time t, measured from a fixed point:

$$m\frac{\mathrm{d}^2 x}{\mathrm{d}t^2} + r\frac{\mathrm{d}x}{\mathrm{d}t} + kx = 0, \qquad (5.2)$$

where r is the damper constant and k is a constant representing the stiffness of the spring. If we examine the dimensions of this equation we find that

$$\frac{\mathrm{ML}}{\mathrm{T}^2} + [r]\frac{\mathrm{L}}{\mathrm{T}} + [k]\mathrm{L} = 0.$$

For dimensional consistency, we see that r must have dimensions MT^{-1} and k must have dimensions MT^{-2}. Alternatively, remembering that MLT^{-2} has the dimensions of force for which the unit is the newton (N), we can see that r can be measured in units of newton seconds per metre ($\mathrm{N\ s\ m}^{-1}$) and k can be measured in units of newtons per metre ($\mathrm{N\ m}^{-1}$) and these are in fact the usual units for these constants.

In our model as represented by equation (5.2), we have two variables x and t, and three parameters m, r and k. We can simplify the equation by making the variables dimensionless and at the same time this reduces the number of parameters.

Let $x = aX$ and $t = bT$, where X and T are our new dimensionless variables and a and b are parameters. Clearly, a has to be a length and b has to be a time. We could choose a to be the value of x when $t = 0$, i.e. $a = x(0)$. What shall we choose for b? In terms of our new variables, we have

$$\frac{\mathrm{d}x}{\mathrm{d}t} = \frac{\mathrm{d}(aX)}{\mathrm{d}(bT)} = \frac{a\,\mathrm{d}X}{b\,\mathrm{d}T}$$

and

$$\frac{\mathrm{d}^2x}{\mathrm{d}t^2} = \frac{\mathrm{d}}{\mathrm{d}t}\left(\frac{\mathrm{d}x}{\mathrm{d}t}\right) = \frac{\mathrm{d}}{\mathrm{d}t}\left(\frac{a\,\mathrm{d}X}{b\,\mathrm{d}T}\right) = \frac{a}{b}\frac{\mathrm{d}}{\mathrm{d}(bT)}\left(\frac{\mathrm{d}X}{\mathrm{d}T}\right) = \frac{a}{b^2}\frac{\mathrm{d}^2X}{\mathrm{d}T^2}.$$

So equation (5.2) becomes

$$\frac{ma}{b^2}\frac{\mathrm{d}^2X}{\mathrm{d}T^2} + \frac{ra}{b}\frac{\mathrm{d}X}{\mathrm{d}T} + kaX = 0$$

or

$$\frac{\mathrm{d}^2X}{\mathrm{d}T^2} + \frac{br}{m}\frac{\mathrm{d}X}{\mathrm{d}T} + \frac{kb^2}{m}X = 0.$$

We have two choices.

1 Take $br/m = 1$.
2 Take $b^2k/m = 1$.

We shall choose the second; so $b = \sqrt{m/k}$ and our equation is

$$\frac{\mathrm{d}^2X}{\mathrm{d}T^2} + \alpha\frac{\mathrm{d}X}{\mathrm{d}T} + X = 0 \tag{5.3}$$

where $\alpha = r/\sqrt{mk}$ (this is often called the 'damping coefficient'). The new equation (5.3) involves two variables X and T and one parameter α. We can produce solutions for $X(T)$ for various values of the parameter α. For a particular problem, we can translate back into $x(t)$ using $x(t) = x(0)X(T)$ and $T = \sqrt{k/m}t$.

5.9 SUMMARY

In this chapter, we have looked briefly at some of the more important skills involved in mathematical modelling. It should not be regarded as a complete list but as an introductory guide.

1 Early on in the modelling process, we need to make a list of *factors*. These include variables, parameters and constants, some of which can be linked together in groups. It is probably a good idea to make the list longer than

necessary to begin with, just to be sure that you have thought of everything. Go through your list, crossing out factors which you think have only a marginal relevance to the problem. Remember that your first model should be as simple as possible. Try to identify *inputs* and *outputs*. Devise a suitable notation and keep track of the *units*.

2 Think carefully about your assumptions and try to make a complete and accurate list.
3 Try to identify which variables affect each other and how the relationship can be represented mathematically.
4 Care is needed when translating verbal statements into mathematical equations. The verbal statements are often vague so that there are a number of mathematical options. Choose simple forms in preference to complicated ones.
5 Compare the relative sizes of the terms appearing in your equations (using data if necessary). This can lead to simplifications without loss of accuracy and also helps to pinpoint areas where more detail is required.
6 Reduce the number of parameters if you can. This ultimately decreases the amount of computation required.

6 USING DATA

6.1 INTRODUCTION

It is possible to build our mathematical models out of the abstract concepts of mathematics using simply pen and paper or blackboard and chalk. During the model-building activity, we are in a dream world of idealisations and imagination but the real world is still somewhere in the background. If our models are to be more than just curiosities of interest only to mathematicians, they must be made to confront reality and it is through data or conclusions made from data that this confrontation occurs.

By data we mean any facts, measurements or observations which have been collected in the real world. They may well be inaccurate and imperfect but, as far as we can tell, they represent the truth with which our models have to be compared. Interaction between models and data occurs in a number of ways, amongst which are the following.

1 Data can be useful in *suggesting* models or parts of models during the development stage when we are trying to form our ideas. Some models (referred to as 'empirical') can even be based entirely on data. Examples of simple empirical models are given in section 6.3.
2 Data are needed to *estimate* the values of parameters appearing in a model. This process of estimating particular parameter values for a particular application is sometimes called 'calibrating' a model. In section 6.4, we look at some of the methods that can be used and some of the problems involved.
3 Data are needed to *test* a model, i.e. to check whether our model's predictions correspond reasonably well to what is observed in the real world. We may also wish to choose the best out of a number of alternative models.

6.2 DATA COLLECTION

When you are given a modelling problem, there may be some data that go with it and in some cases you may have to use what you are given and no more. Very often, however, there is the possibility or the need to obtain more data. The task of collecting such data almost invariably turns out to be harder than you think. The questions that you have to try and answer are as follows.

1 Exactly what data are needed? You will probably have to proceed well into the modelling process before you are able to decide what are the relevant data for your particular problem. It is quite possible that you already have more data than you need and you will have to exercise your judgement in throwing away irrelevant or redundant information.
2 How can the relevant data be obtained? You may have to go back to the person or persons who gave you the problem and ask for the data that you need. The data may already be available or it may be necessary to do further experiments. Another common possibility is that your local or college library will be able to guide you to published sources containing the data that you need. In some cases, you may have to collect the data yourself and, depending on the type of model, this may involve you in activities such as designing questionnaires for statistical surveys or making scientific measurements.
3 In what form do you need the data? If there is a large volume of data, you may want to reduce it by carrying out a statistical summary involving, for example, calculations of means, standard deviations, percentages and histograms. Do not forget to make a note of your sources of data, including dates. These will be needed when you come to write up your report (see chapter 9).

EXERCISES 6.2

For the following exercises, you will need to decide

(a) exactly what data are needed (the problem statement may be quite vague),
(b) how you will obtain the relevant data and
(c) how you will *present* the data (graphs? tables?).

1 Where do students live? In what type of accommodation?

2 How do students travel to college? How long do they take to travel?

3 On what do students spend their money?

4 What kinds of shop are there in the main street where you live?

5 Choose a convenient cross-roads or junction controlled by traffic signals and find how many cars get through when the lights are green.

6 Put n cups of cold water into an electric kettle and find the time $t(n)$ that it takes to boil. Plot a graph of $t(n)$ against n. Can you devise a mathematical model which fits the data?

6.3 EMPIRICAL MODELS

An empirical model is one which is derived from and based entirely on data. In such a model, relationships between variables are derived by looking at the available data on the variables and selecting a mathematical form which is a compromise between accuracy of fit and simplicity of mathematics. It will always be possible to arrange a perfect fit, if necessary, by using a sufficiently complicated mathematical formula, but this is hardly a sensible approach. What is usually required is the simplest formula which will give an *adequate* fit. The important distinction is that empirical models are *not* derived from assumptions concerning the relationships between variables, and they are *not* based on physical laws or principles. Quite often, empirical models are used as 'submodels' or parts of a more complicated model. When we have no principles to guide us and no obvious assumptions suggest themselves, we may (with justification) turn to the data to find how some of our variables are related.

The first step in deriving an empirical model is to plot the data on graph paper. The simplest case will be when all the points lie on or near a straight line. If we know that the data are subject to measurement errors or if a random influence is known to be at work, we may accept a scatter of points around the straight line. In this case, we can either use our own judgement to fit the 'best' line or use the statistical method of least squares to obtain the regression equation in the form $y = ax + b$. Do *not* use the statistical technique automatically; it may be quicker and just as useful to do a fit 'by eye'. A common mistake is to use the form $y = ax + b$ and to employ a standard computer package or calculator routine to find the least-squares estimates of a and b when a little thought shows that y must be zero when x is zero, in which case the form $y = ax$ is required. If the graph shows quite clearly that the relationship between the two variables is not linear, try to plot one or both variables as logarithms, the idea being to get a graph which is reasonably straight.

Example 6.3.1

Table 6.1 records the heights and weights of a sample of 15 people of various ages. Figure 6.1 shows the result of plotting weight against height and the

Table 6.1

Height H/m	0.75	0.86	0.95	1.08	1.12	1.26	1.35	
Weight W/kg	10.0	12.0	15.0	17.0	20.0	27.0	35.0	
Height H/m	1.51	1.55	1.60	1.63	1.67	1.71	1.78	1.85
Weight W/kg	41.0	48.0	50.0	51.0	54.0	59.0	66.0	75.0

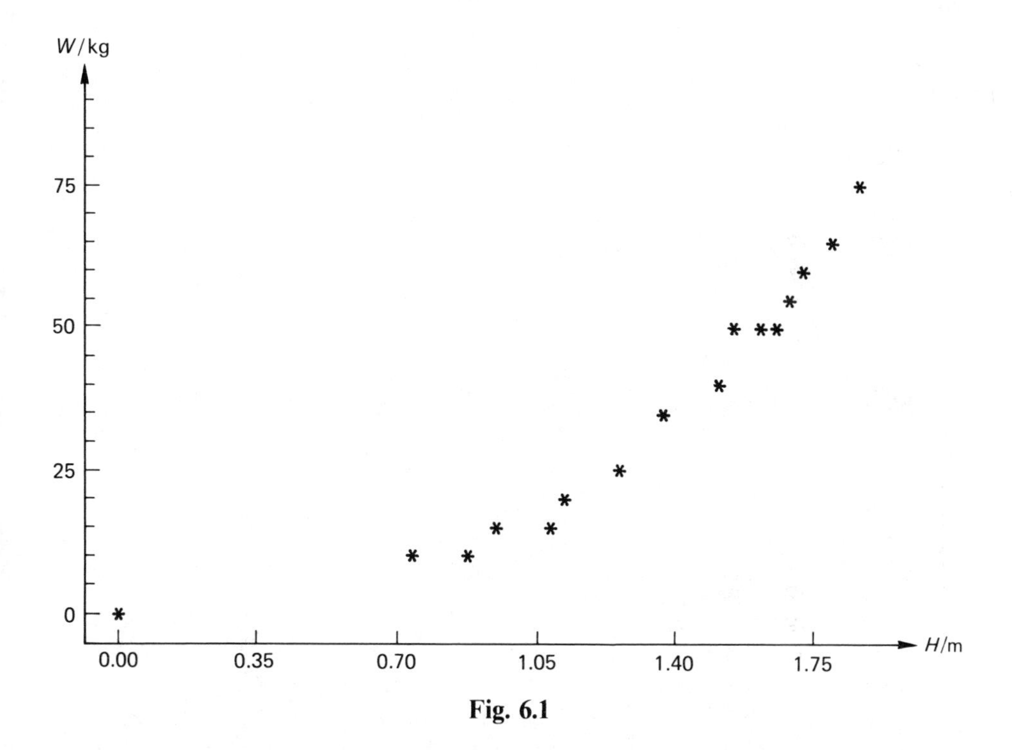

Fig. 6.1

points seem to lie close to some curve. Note that the point (0, 0) is included since zero height must imply zero weight! Both variables have been plotted as logarithms in Fig. 6.2, where $x = \ln H$ and $y = \ln W$. (This time the point $W = 0$, $H = 0$, must be left out. Why?)

The straight line fitted by eye gives

$$y = 2.32x + 2.84.$$

Least-squares fitting of the regression model $y = ax + b$ gives

$$y = 2.30x + 2.82,$$

or

$$\ln W = 2.30 \ln H + 2.82.$$

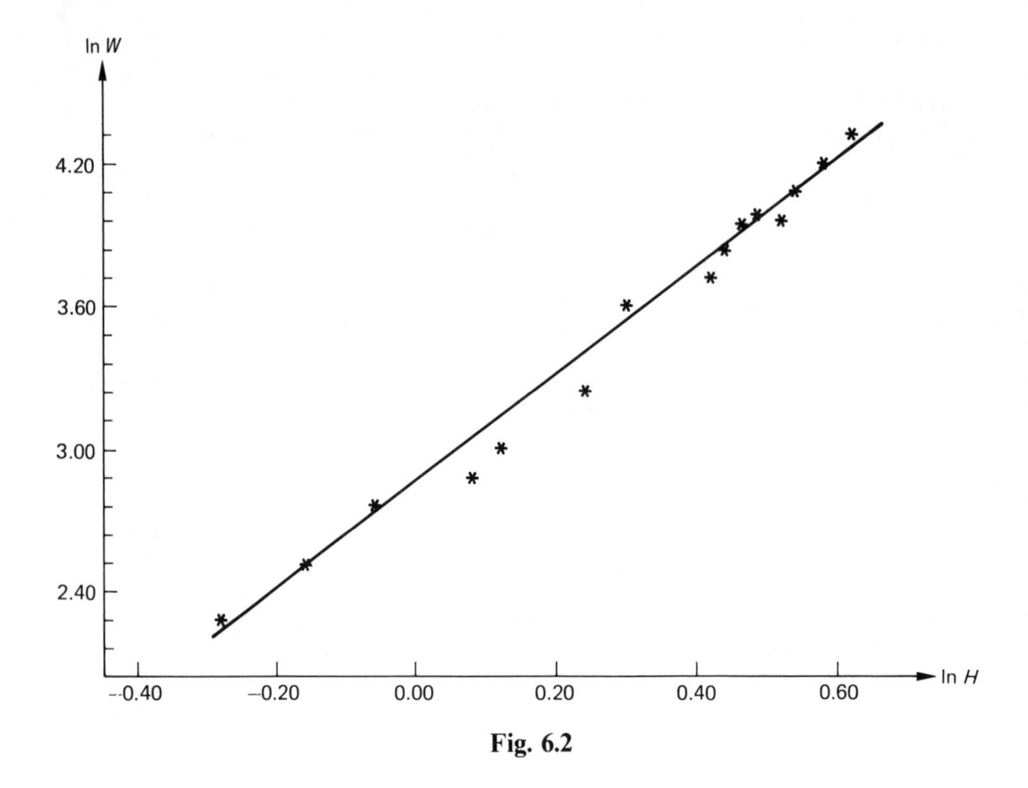

Fig. 6.2

So

$$W = 16.78H^{2.3}.$$

The rather awkward exponent is typical of empirical models. As another example, the drag force on a particle of diameter d moving with speed u relative to a fluid of density ρ and viscosity μ is usually modelled by $F = 0.5C_D u^2 A$, where A is the cross-sectional area of the particle at right angles to the motion. The 'drag coefficient' C_D is given by the following empirical model:

$$C_D = \frac{24}{(Re)}(1 + 0.125(Re)^{0.72}) \qquad \text{for } 0 < (Re) \leqslant 1000,$$

where $(Re) = \rho u d/\mu$ is the particle Reynolds number.

Example 6.3.2

Table 6.2 gives the tide level (in metres relative to a mark on a sea wall) observed over an interval of 2 days. Can we use the data to predict the tide level at 1.00 pm (13:00 hours) on Saturday, 5 December?

Table 6.2

Time	00:00	01:00	02:00	03:00	04:00	05:00	06:00	07:00
Tide level/m, Tuesday, 1 December	2.4	1.2	−0.1	−1.5	−2.5	−3.0	−2.7	−1.6
Tide level/m, Wednesday, 2 December	3.1	2.0	0.6	−0.9	−2.2	−3.0	−3.2	−2.5

Time	08:00	09:00	10:00	11:00	12:00	13:00	14:00	15:00
Tide level/m, Tuesday, 1 December	0.2	2.1	3.4	3.6	2.9	1.6	0.2	−1.2
Tide level/m, Wednesday, 2 December	−0.9	1.1	2.9	3.9	3.6	2.5	1.0	−0.5

Time	16:00	17:00	18:00	19:00	20:00	21:00	22:00	23:00
Tide level/m, Tuesday, 1 December	−2.4	−3.0	−3.1	−2.3	−0.7	1.3	2.9	3.6
Tide level/m, Wednesday, 2 December	−2.0	−3.0	−3.4	−3.0	−1.7	0.2	2.2	3.5

Solution

Our first step is to plot the data to get a visual impression of how the level changes with time. This is done in Fig. 6.3 and it is clear that for this example a straight line or a power function is not the right kind of model. What we need is a function which repeats itself and the obvious choice is something

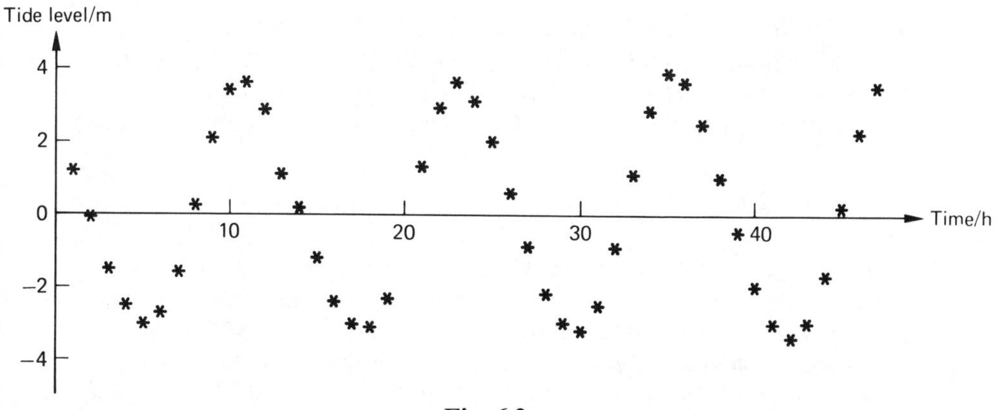

Fig. 6.3

like $x = a\sin(bt)$. From the graph, we measure the period (time of one repetition) to be about 12.3 h; so we need $2\pi/b = 12.3$ h so that $b \approx 0.511\ \mathrm{h}^{-1}$.

The difference between high tide and low tide seems to be about 6.6 m, so that the amplitude is 3.3 m, but our model cannot be just $x = 3.3\sin(0.511t)$ because this gives $x = 0$ at $t = 0$ (taking $t = 0$ to be 00:00 hours on 1 December).

We can either try to estimate the time t_* when $x = 0$ and then use $x = 3.3\sin[0.511(t - t_*)]$ or, equivalently and more conveniently, fit a model of the form

$$x(t) = a\sin(0.511t) + c\cos(0.511t)$$

where $x(0) = c = 2.4$ and $x(23) = a\sin(11.753) + c\cos(11.753) = 3.6$, so that $a = -2.7$. Our model is therefore

$$x(t) = 2.4\cos(0.511t) - 2.7\sin(0.511t).$$

The next step would be to check the values of x given by this model for values of t from 0 to 47 against the data. However, we shall pass over this step and go on to make our prediction.

The point 1.00 pm on 5 December corresponds to $t = 4 \times 24 + 13 = 109$ and so $bt = 0.511 \times 109 \approx 55.7$. Subtracting multiples of 2π ($8 \times 2\pi \approx 50.27$), the answer we are looking for is

$$x(109) = 2.4\cos(5.43) - 2.7\sin(5.43) \approx 3.6.$$

The *actual* value observed was 4.1; so our model's prediction was in error by 0.5/4.1 or about 12%.

Note that we have used the model to *extrapolate*, i.e. to make a prediction for a value of t outside the range of the data. This is a risky thing to do and it is better and safer to use a model only over the range of data from which it was derived.

Further thoughts

We notice from the graph that the mark $x = 0$ does not seem to be quite midway between high and low tides and that the amplitude of the oscillations seems to be increasing slightly with time. How can we modify our model to take account of these details? Would we then get a more accurate prediction?

Example 6.3.3

Sports and games provide an area where data are in plentiful supply, nowhere more so than in track and field athletics. Over the years, athletes' performances have gradually improved and there seems to be no reason why performances should not continue to improve in the future. It is interesting to see what mathematical modelling can do here; so we shall examine an example from track athletics. The following data give times for the men's

400 m (see S. Greenberg, *The Guinness Book of Sporting Facts*, Guinness Superlatives Ltd, 1982).

Year Y	1910	1924	1933	1948	1960	1980
Time T/s	48.3	47.5	47.0	46.1	45.6	44.7

Now the aim in this chapter is to fit a model to the data which means an appropriate formula is sought to relate the variables Y and T. Such a formula is useful to help to predict future performance, for instance what time can be achieved by the year 2000. The choice of mathematical formula is the first consideration and good advice is first to plot the data on a graph to get the feel of the relation. This is shown in Fig. 6.4. This has an approximately linear appearance; so a possible relation is

$$T = a - bY, \tag{6.1}$$

where a and b are positive constants. A straight line can be drawn through the points *by hand* and from it values of a and b obtained.

Taking the coordinates of the points A and B from the graph gives A (1920, 47.75) and B (1970, 45.125). Substituting these values into equation (6.1) gives two equations for the determination of a and b:

$$47.75 \ \ = a - 1920b,$$

$$45.175 = a - 1970b.$$

Solving these, $a = 146.63$ and $b = 0.0515$. The resulting formula is therefore

$$T = 146.63 - 0.0515Y. \tag{6.2}$$

Some judgement has been used in drawing the line and the question arises as to whether we have drawn the *best* line according to some test. This is a well-known problem in statistics and is referred to in the next section. We shall concentrate here on the choice of formula and now think about the validity of the linear relation obtained. The validity can be checked by using equation (6.2) to predict T values in the future. For $Y = 2000$, substituting into equation (6.2) gives $T = 43.63$. On the other hand, by the year 2200, equation (6.2) gives a time of 33.33 s for the race which is a little dubious, even though it is difficult to image what conditions will be like so far ahead.

We may argue that the range of Y values of interest is restricted by $1900 < Y < 2000$, in which case the linear relation is acceptable. However, if a closer inspection is made of Fig. 6.4, you can see that the data points lie on a curve which is concave upwards; so perhaps there is an alternative formula which represents the data more accurately. Common sense suggests

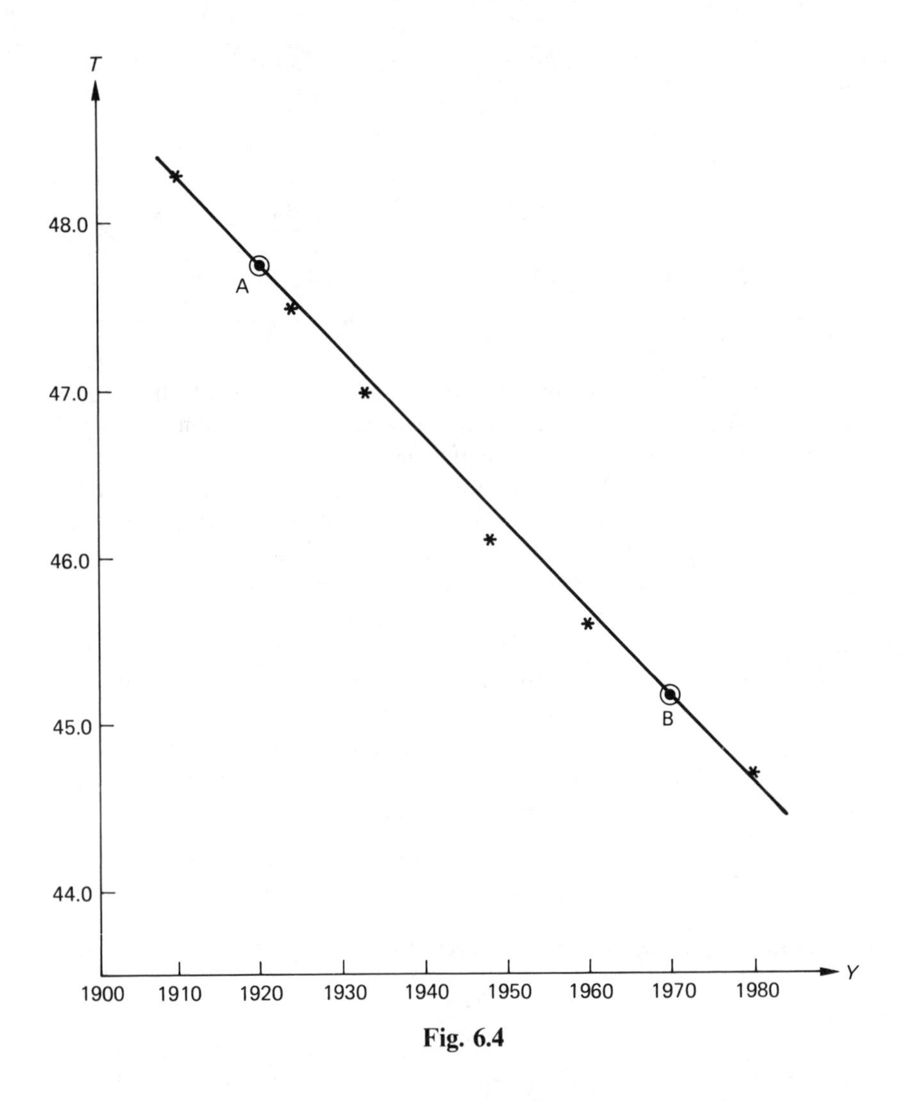

Fig. 6.4

that there must be some physical limit beyond which an athlete cannot go; T must have some minimum value, perhaps around 35 s.

Also, looking at the background to the data, an attempt can be made to explain the trend shown by Fig. 6.4. Over the years, health and fitness have improved owing to a number of factors and, in particular, athletes have trained more scientifically. The quality of athletics equipment and track surface has also contributed to improving times in races such as the 400 m.

The data show that, to a slight extent, improvement is now tailing off and this may be explained by claiming that improvements due to the above factors have now largely been made. In the future the record time for 400 m will only be lowered by small amounts achieved by increased competition and intensive training.

It is possible to argue against this theory of course, and that is one of the features of mathematical modelling. At any rate, following the above ideas, the trend that we want can be represented by the formula

$$T = \alpha \exp(-\beta Y) + \gamma.$$

This means that three parameters have to be calculated. This can be done using the method of least squares (although this results in quite a tricky numerical problem here). In fact, if we drop the parameter γ from the model, then α and β can be calculated by elementary methods.

A further point is that Y can be replaced for convenience by $X = Y - 1900$; so in the end we are using the formula

$$T = \alpha \exp(-\beta X). \tag{6.3}$$

Taking logarithms gives

$$\ln T = \ln \alpha - \beta X. \tag{6.4}$$

Organising the data for a graph plot of X and $\ln T$ gives the following.

X	10	24	33	48	60	80
$\ln T$	3.877	3.861	3.850	3.831	3.820	3.800

(Note here in passing the small variation in the range of $\ln T$ values—we must always examine data for *mathematical* as well as *modelling* suitability. Here the third decimal digit cannot be dropped.)

From the resulting graph shown in Fig. 6.5 a linear relation between X and $\ln T$ is expected. This looks fairly acceptable and the line of best fit is drawn in by hand. The gradient of the line, β, is calculated as equal to $0.055/50 = 0.0011$. The parameter α is found from the intercept on the $\ln T$ axis: $(\ln T)_{X=0} = 3.886$, and so $T_{X=0} = 48.76$. Finally, this model gives the result

$$T = 48.76 \exp[-0.0011(Y - 1900)]. \tag{6.5}$$

Again testing the model, when $Y = 2000$, $T = 43.64$, which is in close agreement with the earlier results. On the other hand, when $Y = 2200$, $T = 35$, which is only slightly more realistic.

The conclusion for this example is that the linear model is adequate to represent the data over the twentieth-century time span, whereas the exponential only shows a significant difference far into the future. We did decide to omit the parameter γ from the exponential model, which will have had some effect.

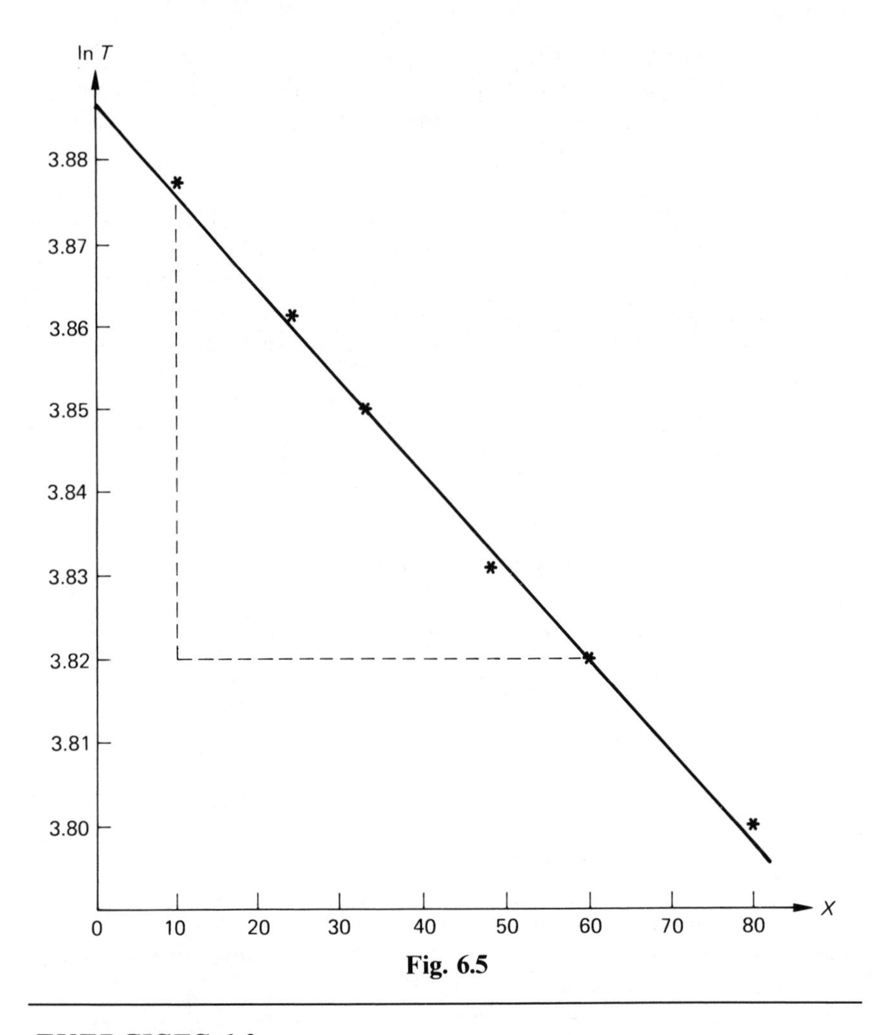

Fig. 6.5

EXERCISES 6.3

1 The distances thrown by the women's shot put winner in the post-war Olympic Games are as follows.

Year Y	1948	1952	1956	1960	1964
Distance D/m	13.75	15.28	16.59	17.32	18.14

Year Y	1968	1972	1976	1980	1984
Distance D/m	19.61	21.03	21.16	22.41	23.57

Plot a graph of the data and examine the trend of performance over the years. Draw a line of best fit and calculate from it the predicted winning distance for the year 2000. Is your result realistic?

2 The marathon race in the Olympic Games is the classic event, run over a distance of 26 miles 385 yd. As an exercise more suitable to chapter 4, calculate how many metres this is. Data for the winning times since the war are as follows.

Venue	Year	Time
London	1948	2 h 34 min 52 s
Helsinki	1952	2 h 23 min 03 s
Melbourne	1956	2 h 25 min 00 s
Rome	1960	2 h 15 min 16 s
Tokyo	1964	2 h 12 min 11 s
Mexico City	1968	2 h 20 min 26 s
Munich	1972	2 h 12 min 20 s
Montreal	1976	2 h 09 min 55 s
Moscow	1980	2 h 11 min 03 s
Los Angeles	1984	2 h 09 min 21 s

Plot a graph of these data to appraise their trend. Explain the result for 1968 if you can. Suggest a mathematical formula to represent the data based on the ultimate winning time of 2.00 h.

3 Soap powder for automatic washing machines is sold in packets of different sizes (i.e. different masses). The following data were collected in one supermarket.

Mass/kg	0.93	3.10	4.65	6.20
Price/£	0.91	2.75	3.99	4.99

Plot the data. Do the points lie on a straight line? If not, can you fit a simple empirical model to the data?

4 High water at Aberystwyth (12–20 August 1987) was as follows.

	Time	Height/m	Time	Height/m
Wednesday, 12 August 1987	10:37	15.7	22:49	17.1
Thursday, 13 August 1987	11:19	14.8	23:33	16.1
Friday, 14 August 1987			12:02	14.1
Saturday, 15 August 1987	00:18	15.1	12:49	13.1
Sunday, 16 August 1987	01:07	13.8	13:41	12.1
Monday, 17 August 1987	02:05	12.5	14:48	11.5
Tuesday, 18 August 1987	03:22	11.5	16:17	11.2
Wednesday, 19 August 1987	04:59	11.2	17:41	11.8
Thursday, 20 August 1987	06:17	11.5	18:40	12.5

Can you use the data to 'predict' the time and height of the evening tide on Monday, 26 August 1987? (The actual height observed was 15.1 m at 20:55.)

5 The following are planetary data.

Position from Sun n	Name of planet	Mean distance from Sun $R/10^9$ m	Period T/days
1	Mercury	57.9	88
2	Venus	108.2	225
3	Earth	149.6	365
4	Mars	227.9	687
5	Asteroid belt	330–490	Various
6	Jupiter	778.3	4 329
7	Saturn	1427	10 753
8	Uranus	2870	30 660
9	Neptune	4497	60 150
10	Pluto	5907	90 670

(a) Obtain an empirical model for finding T from R. (There is a theoretical model based on Newton's mechanics.)

(b) Obtain an empirical model relating n and R.

6.4 ESTIMATING PARAMETERS

As we saw in chapter 5, our main concern in mathematical modelling is to formulate relationships between *variables* but these also almost invariably involve *parameters*. For example,

$$y = ax^2 - bx, \qquad x > 0, a > 0, b > 0,$$

where x and y are variables, and a and b are parameters. In this general form a model can be of some use in predicting *general* behaviour in a descriptive fashion, such as 'when x increases from zero, y increases until $x = b/2a$ and then decreases, becoming zero at $x = b/a$'. To use the model in a direct practical way, however, we must obtain numerical values for the parameters and we can only get these from data. The process of using data to obtain parameter values relevant to a particular application of the model is often called 'calibrating' or 'tuning' the model. The usual methods of obtaining the parameter values are as follows.

1 Graphical.
2 Statistical, usually involving least-squares estimation.
3 Mathematical, usually requiring the solution of linear equations.

Example 6.4.1

In section 6.3, we obtained the empirical model $W = 16.78H^{2.3}$ from the data on 15 people. Without any data, what kind of model could we develop for predicting a person's weight from his or her height?

Solution

A very simple model can be derived by assuming that all human beings are different-sized copies of the same geometric shape and made of the same material. Since weight is proportional to volume (for constant density) and volume is proportional to (height)^3, the simplest model is $W \propto H^3$. This model predicts $W = aH^3$ for some value of a. We cannot say anything about the value of the parameter a without data. It would clearly be sensible to use different values of a for males and females.

If the data in Table 6.1 are all that we have, how do we estimate the value of a? Using the *graphical* approach, we draw the graph of W against H^3 (Fig. 6.6) and fit the best straight line that we can through the points, not forgetting that W must be 0 when $H = 0$; so our line must pass through the point $(0, 0)$. Calculating the slope of the line we find that $a \approx 12.3$.

Using the *statistical* approach, the regression model $y = ax + b$ (where $y = W$ and $x = H$) gives $y = 11.1x + 4.19$. This predicts that a person of zero height will weigh 4.19 kg! Least-squares fitting of the more sensible model $y = ax$ is done by choosing the value of a which minimises the sum of squares of the errors $S = \sum(y - ax)^2$. We have $\mathrm{d}S/\mathrm{d}a = 0$ when $\sum[-2a(y - ax)] = 0$, i.e. $\sum(y - ax) = 0$ or $\sum y = a\sum x$. From the data, $\sum y = 580$ and $\sum x = 46.259$, giving $a \approx 12.54$.

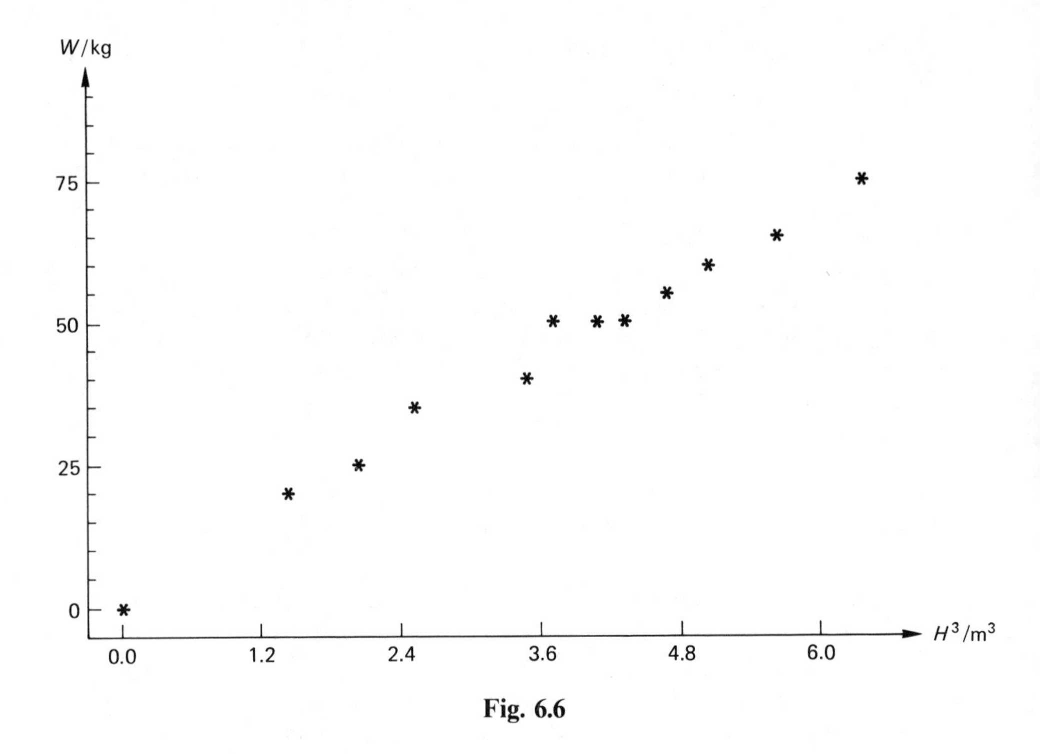

Fig. 6.6

The *mathematical* approach can be used when only a very limited number of data are available. In this case, since we know that the point $(0, 0)$ is on the line, we need only one other point. If we take the point $(6.332, 75)$, then solving $75/6.332 = a$ we get $a \approx 11.84$.

Of the three estimates of a, the one which is likely to be most reliable is the statistical estimate $a = 12.54$, but with such a small sample we should not expect our model to give accurate predictions of the weights of other people from their heights.

It is interesting to note that the quantity W/H^2 is used by medical researchers as a measure of obesity. With W measured in kilograms and H in metres the value $W/H^2 = 27$ is taken as the norm.

Example 6.4.2

On our video recorder, there is a four-digit display which can be set to 0000 at the beginning of a tape. At the end of a '180 min' tape the display reads 1849, while the actual elapsed time was measured to be 185 min 20 s. We noticed that it took 3 min 21 s for the display to change from 0084 to 0147. At present, we have a recording which ends at 1428. Have we got enough room left on the tape to record a programme 60 min long?

Solution

To answer this question, we shall develop a model which will enable us to write down a formula relating the counter reading to elapsed time. The objective for our model is therefore quite clear: *given* the counter reading n, *find* the elapsed time t. We shall develop the model quickly rather than go through the modelling process in detail.

We imagine the tape to be of constant thickness w and wound around a circular take-up spool of radius r. To complete the model we need to make the following two assumptions.

1 The tape moves at a constant linear speed v across the read–write heads.
2 The reading on the display is proportional to the number of revolutions of the take-up spool.

Suppose that, after the tape has been running for a time t, the take-up reel looks like Fig. 6.7. The total area of tape (viewed from the side) is $\pi(R^2 - r^2)$, where R is the outer radius. Note that R and l are variables and r is a

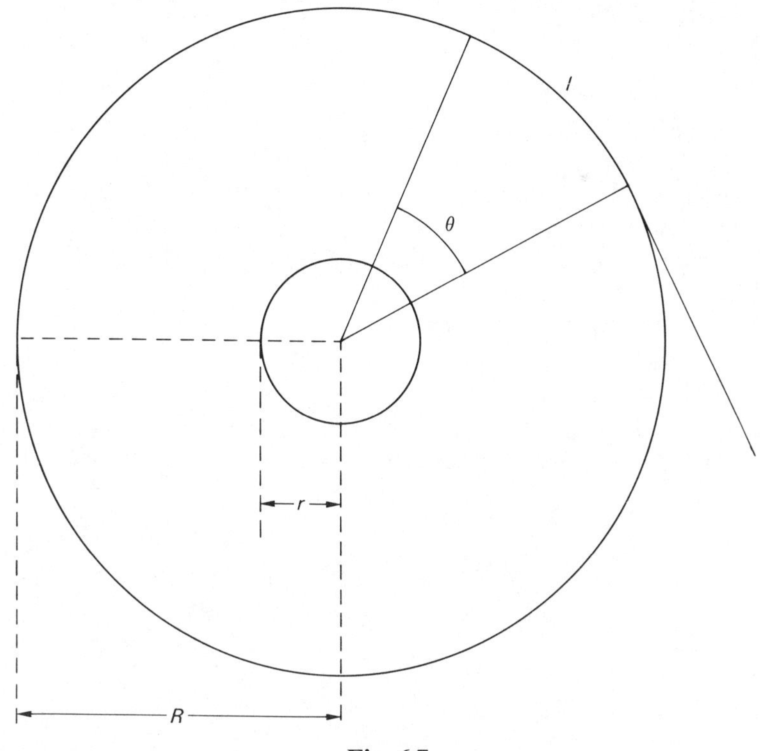

Fig. 6.7

parameter. The length of tape which has been taken up is given by the area divided by the width, i.e. $\pi(R^2 - r^2)/w$. It is also equal to vt by assumption 1; so

$$R = \left(\frac{wvt}{\pi} + r^2\right)^{1/2}.$$

When the take-up spool turns through a small angle $\Delta\theta$ rad, it picks up a length $\Delta l = R\,\Delta\theta$ of tape; so $\Delta\theta = \Delta l/R$ and $\Delta l = v\,\Delta t$. Therefore,

$$\mathrm{d}\theta = \frac{v\,\mathrm{d}t}{R} = v\left(\frac{wvt}{\pi} + r^2\right)^{-1/2}\mathrm{d}t,$$

and so

$$\int_0^\theta \mathrm{d}\theta = \int_0^t v\left(\frac{wvt}{\pi} + r^2\right)^{-1/2}\mathrm{d}t$$

and

$$\theta = \left[\frac{2\pi}{w}\left(\frac{wvt}{\pi} + r^2\right)^{1/2}\right]_0^t.$$

By assumption 2, $n = k\theta$; so we can write

$$n = \frac{2\pi k}{w}\left[\left(\frac{wvt}{\pi} + r^2\right)^{1/2} - r\right].$$

We now have a choice of two directions. We could try to get numerical values for w, v and r, possibly from the manufacturers, and assume a value for k (say $k = 1/2\pi$). We shall not take this course here. The alternative is to simplify the expression so that it involves as few parameters as possible (see section 5.9 for ways of doing this) and then use the data to get values for the parameters.

The expression for n is equivalent to

$$n = \alpha(\sqrt{t+\beta} - \sqrt{\beta}),$$

where $\alpha = 2k\sqrt{\pi v/w}$ and $\beta = r^2\pi/wv$. This reduces the number of parameters to two and we can go no further. The data that we have available are as follows.

t/min	n
0	0
185.33	1849
Unknown, t_1 say	0084
$t_1 + 3.35$	0147

The $n = 0$, $t = 0$ condition was built into our model (when we wrote down the limits of integration). The other items of data lead to the equations

$$1849 = \alpha(\sqrt{185.33 + \beta} - \sqrt{\beta})$$

$$0084 = \alpha(\sqrt{t_1 + \beta} - \sqrt{\beta})$$

$$0147 = \alpha(\sqrt{t_1 + 3.35 + \beta} - \sqrt{\beta})$$

We have three equations for three unknowns α, β and t_1, but they are very awkward to solve. Is there a better way?

Remembering that our objective is to find t from n, we rewrite the equation $n = \alpha(\sqrt{t + \beta} - \sqrt{\beta})$ as $\sqrt{t + \beta} = n/\alpha + \sqrt{\beta}$ so that $t + \beta = (n/\alpha + \sqrt{\beta})^2$, leading to an equation of the form

$$t = an^2 + bn,$$

where a and b are parameters.

Substituting the data values, we have

$$185.33 = (1849)^2 a + (1849)b, \tag{6.6}$$

$$t_1 = (84)^2 a + (84)b, \tag{6.7}$$

$$t_1 + 3.35 = (147)^2 a + (147)b. \tag{6.8}$$

It is now easy to get rid of t_1 (which we do not really want to know) by subtracting equation (6.7) from equation (6.8). Solving the resulting equation together with equation (6.6), we find $a = 0.000\,029\,1$ and $b = 0.046\,46$ and our model is

$$t = 0.000\,029\,1n^2 + 0.046\,46n.$$

Noticing that a is substantially smaller than b, we wonder whether we could leave the an^2 term out of the model which would then be nice and simple, $t = 0.046\,46n$. On second thoughts, we realise that with $n \sim O(10^3)$ the an^2 term is $\sim O(10^6 \times 10^{-5}) \sim O(10)$ while $bn \sim O(10^3 \times 10^{-2}) \sim O(10)$; so the two terms are about the same order of magnitude. We must keep them both.

Note that we have carried out a mathematical fit here rather than a statistical fit, simply because of the shortage of data. The number of data that we have are the bare minimum that we need to find values for the two parameters. It would be better if we had data on n and t values at a number of points. A statistical fit would then give us more reliable estimates for a and b.

We can now use our model to answer the original problem. At $n = 1428$ the corresponding value for t from our model is

$$t = 0.000\,029\,1 \times (1428)^2 + 0.046\,46 \times 1428 = 125.69.$$

The amount of recording time left on the tape is $(185.33 - 125.69)$ min =

59.64 min; so there is not quite enough time for recording the 60 min programme (according to our model).

EXERCISES 6.4

1 The following data were collected from the same video tape recorder as featured in Example 6.4.2. How well does the model fit the data?

Counter reading n	0	200	400	600	800
Elapsed time t/min	0	10.67	23.57	38.80	56.25
Counter reading n	1000	1200	1400	1600	1800
Elapsed time t/min	75.94	97.76	122.28	148.87	178.7

2 Consider the following alternative models. If we calculate the average time of one revolution of the take-up spool, we get $(185 \times 60 + 20)/1849 \approx 6.014$ s. We could get a rough answer by assuming that the take-up spool rotates at this constant average speed. How good is the answer obtained?

It is clear that the spool does not rotate at a constant rate. It is rotating quickly at the beginning and slower at the end when it carries a large amount of tape. From our original data, the average time of one revolution over the interval from counter reading 0084 to 0147 was $(3 \times 60 + 21)/(147 - 84) \approx 3.01$ s. Since the rate of revolution is *not* constant, what is the simplest model that we can set up for it? Clearly this is a straight line of the form $A + Bn$, where n is the counter reading. Use the data to find values for A and B and use this model to answer the original question. How good is the answer obtained?

6.5 ERRORS AND ACCURACY

Our data very often come from experimental observations and measurements made with imperfect instruments; so we cannot hope to avoid errors. The errors in the data are of course in addition to the errors that we make during the modelling. We can list the various sources as follows.

1 Errors due to our modelling assumptions.
2 Errors due to using an approximate method of solution.
3 Errors due to carrying out inexact arithmetic (rounding errors).
4 Errors in the data.

With all these errors about, it is clear that our modelling predictions cannot be 100% reliable. We have a duty, however, to try to estimate what the *maximum error* in our prediction is likely to be. In other words, we ought to try to produce a statement of the form 'we predict X with a possible maximum error of Y (or Z%)'.

Errors can be described in more than one way. We define *absolute error* to be the difference between a *true value* and an *approximate value*. For example, if we use 22/7 as an approximation for π the absolute error is $\pi - 22/7 \approx -0.0012645$. An absolute error of 0.1 with respect to a measurement of magnitude 1000 represents greater accuracy than an absolute error of 0.1 in a measurement of magnitude 10. To allow for this distinction, we use *relative error*, defined by relative error = (absolute error)/measurement. It is often expressed as a percentage. For example, the relative error in 22/7 when used as an approximation for π is $0.0012645/\pi \approx 0.0004025$ or about 0.04%.

Let us now look at the above-listed sources of error.

1 It is almost impossible to be sure about the effects of errors due to our modelling assumptions because we do not normally know the correct value of the answer that we are trying to estimate. We can do a certain amount of investigation by varying the assumptions and noting the effect on the answer (see section 6.7).
2 We often use approximate (numerical) methods, either because mathematical solutions are impossible to obtain or just because it is quicker to use a computer. These numerical methods involve errors (truncation errors) depending partly on the data and partly on the type of method used. Sometimes, we use shorter or easier mathematical expressions in place of others such as when we replace $\sin x$ by x and $\cos x$ by 1 when x is small (see section 5.7). This also involves us in truncation errors.
3 Any numerical work that we carry out using an electronic aid such as a calculator or computer inevitably involves rounding errors because of finite storage capacities. These rounding errors can accumulate to an appreciable amount if we do a large amount of computation.
4 Ideally, an estimate of the maximum error in the data should be available.

The way in which all these errors combine together is complicated. As modellers, we should try to identify the most serious source of error and also to estimate the worst possible error in our final prediction.

Example 6.5.1

In chapter 4, we mentioned the formula for translating from temperatures measured in degrees Celsius to the equivalent temperatures in degrees Fahrenheit:

$$°\mathrm{F} = \frac{9}{5} \times °\mathrm{C} + 32.$$

A simpler model would be $°\mathrm{F} = 2 \times °\mathrm{C} + 30$ and this is simple enough for us to do the calculation in our heads. For example 20 °C becomes 70 °F according to the simple model, while the exact formula gives 68 °F. How accurate is this simple formula over the range of air temperatures normally experienced in this country?

Solution

The absolute error is

$$\left(\frac{9}{5} \times °\mathrm{C} + 32\right) - (2 \times °\mathrm{C} + 30) = -0.2 \times °\mathrm{C} + 2.$$

For the range $-5 < °\mathrm{C} < 25$ the error varies from 3 to -3 with a maximum percentage error of about 22% at $°\mathrm{C} = -5$. If we restrict the model to positive temperatures, the error varies from about 5% at 0 °C to about 10% at 25 °C.

Remember that there is no practical advantage in having a very accurate model if an approximate model is easier and quicker to use and gives answers that are sufficiently *accurate* for practical use.

EXERCISES 6.5

1 The function $\sinh x$ is defined by $\sinh x = [\exp x - \exp(-x)]/2$. As x increases, $\exp x$ increases while $\exp(-x)$ decreases; so we can expect that, for sufficiently large x, $\sinh x$ can be approximated by $(\exp x)/2$. How large does x have to be for this to be accurate to less than 5%?

2 When the Earth is modelled as a sphere (of diameter 12.72×10^3 km), it is obvious that, if the walls of a tall building are vertical, they cannot be parallel. Suppose that a tower block is 400 m tall and that the ground floor has an area of 2500 m^2. How much extra area is there on the top floor?

3 A railway line is exactly 1 mile long and one night a vandal cuts the line in the middle and inserts an extra foot of metal. The line is constrained at its two ends; so it is now forced to bend into an arc of a circle. How high above the ground is it at the midpoint?

4 In Fig. 6.8, C is the midpoint of an arc AB of a circle. An estimate of the length of the arc is given by $(8b - a)/3$, where a and b are the lengths of AC and AB, respectively. How accurate is this estimate? You may like to

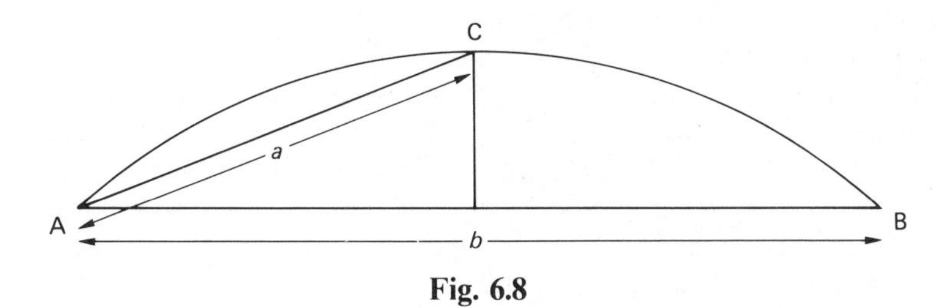

Fig. 6.8

show that, if l is the required length of arc and r is the radius of the circle, then

$$\frac{8b-a}{3} = \frac{1}{3}\left[16r\sin\left(\frac{l}{4r}\right) - 2r\sin\left(\frac{l}{2r}\right)\right].$$

The approximation $l \approx (8b - a)/3$ can be obtained from this by using series approximations for the sin terms.

6.6 TESTING MODELS

The success of a model can be measured in terms of how well its predictions compare with what is actually observed in the real world. In the previous two sections, we have discussed the procedure of *making* our model fit available data. The real test of a model comes when we use it to make a *prediction* for data which we do not have.

By 'testing' a model, we do not mean finding out whether it is 'correct'. In a sense, it is bound to be 'wrong' because it is only a model and has been derived by making simplifying assumptions. When we test a model, what we are trying to find out is whether it gives predictions which are sufficiently accurate to be useful.

In particular, does it adequately fulfil the purpose for which it was designed? Remember always that a model should give *useful* predictions that can be directly applied to the real world. Suppose that you have worked out a clever model for the game of snooker which can produce a prediction such as 'to pot the black ball into the top right-hand pocket, the cue ball should be given an initial velocity of 1.5 m s^{-1} at an angle of 52° to the top cushion'. This *could* be, scientifically speaking, a very accurate prediction but of what practical use would it be to the player who is about to pick up his cue and play the shot? A good model should provide direct *practical* advice.

In Example 6.4.2, our model's prediction was that there would be 59.64 min of recording time left on the tape. When we tested it, we found in fact that 59.42 min were left.

This model therefore served its purpose, but in general we expect our models to do more than to predict a single answer. We should be able to derive both qualitative and quantitative statements from our model, e.g. 'y increases as x increases' or 'y is proportional to x'.

The simplest way to test these statements (assuming some data are available) is to draw graphs. The simple model of Example 6.4.1 predicts that people's weights will be proportional to the cubes of their heights. We can use the data to test this prediction but instead of plotting weight against height, which our model predicts will be a cubic curve, a better idea is to plot weight against $(\text{height})^3$. According to the model, this graph should be a straight line and departure from a straight line is easier to assess. The plot of $\ln H$ against $\ln W$, of course, serves the same purpose.

Not surprisingly (Fig. 6.6) the points do not lie exactly on a straight line. We have to decide whether they are close enough for the model to be useful. Note that we do *not* need to know the value of the parameter. We are testing to see whether the points lie on a straight line and not on some particular line.

When testing any model, try to derive a relationship which your model predicts will be a straight line; then plot the corresponding data on graph paper. Ideally, you should decide what kind of test you are going to do and *then* collect the relevant data.

When the testing (or, as it is often called, the 'validation' or 'verification') of your model has proved disappointing, your task as a modeller is to decide whether to go back to the formulation stage and to improve the model or to scrap it altogether and to start again. Hopefully, the less drastic course of action is chosen. How do you improve your model?

Look at the list of factors again. Have you omitted one which ought to have been included? Have you lumped together some factors which should be considered separately? Look at your assumptions again. Can they be relaxed or made more general? For example, where you have assumed a linear relationship, perhaps a non-linear one is more appropriate?

If you find that your model's predictions are sufficiently accurate but not perfect, try to assess the accuracy. As we saw in section 6.5, it is very important to give a *bound* (i.e. an upper limit) on the possible error. For example, the simple height–weight model $W = 12.54H^3$ gave predictions which were within 47% of the correct values (the worst error).

If you have a reasonable number of data, calculate the standard deviation of the errors in your model's predictions so that you can give an approximate confidence interval for future predictions. Also remember to allow for the *sensitivity* of your predictions to changes in the parameter values and/or changes in the modelling assumptions. Make small changes to your model and find whether or not these give rise to large changes in the model's predictions.

Example 6.6.1

A stone is dropped down a vertical shaft in order to estimate the depth of the shaft by listening for the noise as the stone strikes the bottom. A student develops the following model relating time t and height h:

$$h=\frac{g}{k}\left(t+\frac{1}{k}\exp(-kt)\right)-\frac{g}{k^2},$$

by assuming that the force on the stone due to air resistance is directly proportional to the speed of the stone, k being the proportionality constant.

If air resistance is neglected, the appropriate model is $h=0.5gt^2$ (motion under constant acceleration g). This should agree with the student's model in the case $k=0$ and we ought to check this. However, we cannot put $k=0$ directly into the model (division by zero). Instead, if we expand the exponential term using Taylor's series

$$\exp(-x)=1-x+\frac{x^2}{2}-\dots$$

we obtain

$$\exp(-kt)=1-kt+\frac{k^2t^2}{2}+\mathrm{O}(t^3)$$

and

$$h\approx\frac{g}{k}\left[t+\frac{1}{k}\left(1-kt+\frac{k^2t^2}{2}\right)\right]-\frac{g}{k^2},$$

which gives $h=0.5gt^2$ when $k=0$.

In order to use the model to calculate h from t, we need the value of k. The student has been told that the appropriate value of k to use is 0.05 s^{-1}. Is the model dimensionally correct?

For a particular shaft the noise is heard after 4 s. From the model, we calculate h to be

$$\frac{9.81}{0.05}\left(4+\frac{1}{0.05}\exp(-0.2)\right)-\frac{9.81}{0.05^2},$$

which is about 73.50 m. We now have an estimate of h but what if the 'correct' value of k is actually not 0.05; how much does it matter? If we change k by 10% and substitute 0.045, we get $h\approx 73.98$ m; so a 10% uncertainty in k leads to less than 1% error in h and the model is not unduly sensitive to the value of k.

If the student had neglected air resistance in his or her model would the estimate of h have been very inaccurate? The simple model gives $h=0.5\times 9.81\times(4)^2\approx 78.48$ m. So we have an absolute error of about 5 m or a relative error of about 7%. We conclude that the inclusion of the air resistance has a significant effect.

Another assumption that has been made in this model is that the instant at which the stone hits the bottom and the instant that the noise is heard are the same, but we know that the speed of sound is finite (about 330 m s^{-1}). Does this matter?

The correction to the value of t is the time that it takes for the sound to travel up, which is $h/330 \approx 73.5/330 \approx 0.223$ s. Substituting $t = 4 - 0.223 = 3.777$ in the model gives $h \approx 65.77$ m. We conclude that allowing for the finite speed of sound, which the student completely forgot about, is even more important than allowing for the air resistance.

EXERCISE 6.6

1 In sections 6.3 and 6.4, we derived the following two models for predicting weight from height for humans:

$$W = 16.78H^{2.3} \qquad \text{(the empirical model)},$$

$$W = 12.54H^{3} \qquad \text{(simple theoretical model)}.$$

Which model gives the best fit to the data in Table 6.1?

6.7 SUMMARY

Relevant data are needed at several stages in the modelling process.

1 Initially, to help to form our ideas when we are trying to develop a model.
2 To fix the values of *parameters* which we are including in our model.
3 Most importantly, to *test* our model, i.e. to check whether our model's predictions correspond sufficiently well with what is observed in the real world.

Empirical models are derived entirely from data by choosing the simplest mathematical forms that give a good fit.

When testing models, find a relationship which is predicted to be a straight line and plot the corresponding data.

7 USING RANDOM NUMBERS

7.1 INTRODUCTION

In chapter 2, we came across situations where chance played a part and some of the variables were *random* variables. In chapter 3, we described models involving random variables as 'stochastic' or 'probabilistic' models. They obviously form a very important class of problem since uncertainty is an ever-present feature of real life. Here we look a little more closely at what is involved in developing and using probabilistic models.

7.2 MODELLING RANDOM VARIABLES

What exactly do we mean by a random variable? Clearly a random variable is one whose value is unpredictable in advance, but there is more to it than that. Although an individual measurement of a random variable is unpredictable (except possibly to say that the measured value cannot fall outside a certain range) the long-term pattern of values *is* predictable in a statistical sense. We take the view that randomness is not chaos but has a *pattern* to it and it is this pattern of randomness that we try to model.

Can we find a model for the following data?

Time t/s	0	1	2	3	4	5	6	7	8	9
Variable X	1	0	2	2	1	2	0	1	0	2

X appears to be a discrete random variable taking the values 0, 1 and 2. It is not possible to write down a deterministic formula relating X and t. Instead, we describe the pattern of randomness mathematically by counting the number of times that the different values of X occur and putting together a *frequency table* as follows.

X	0	1	2
Frequency	3	3	4
Relative frequency	0.3	0.3	0.4

This table gives us the *pattern* for the random variable X. The relative frequencies tell us what percentages of the measurements of X have resulted in that particular value. If we needed to build a model involving this random variable, we could assume that this pattern will always apply no matter how many times we measure X. We would be making a general assumption based on very few data and in effect treating the relative frequency distribution as a *probability function*. An alternative would be to *assume* a simple theoretical model for the probability function. For example, we could assume that the three values are in principle equally likely so that the probability function is as follows.

x	0	1	2
$p(x)$	$\frac{1}{3}$	$\frac{1}{3}$	$\frac{1}{3}$

This example illustrates the two choices that we have in creating models for random variables.

1 Use a theoretical model. In any book on mathematical statistics, you will find several standard theoretical models such as the binomial distribution and we shall discuss the main ones later. Each one is based on certain assumptions holding true; so our choice depends on making the appropriate assumptions for our particular model.
2 Use the relative frequency table based on actual data without attempting to fit a standard theoretical model.

The advantage of the first approach is that we can make exact statements about the probabilities of various outcomes and we have the power of mathematical statistics at our disposal. This advantage, however, is limited to fairly simple situations. As the complexity increases, the mathematics becomes too difficult. The advantage of the second approach is that we are conforming exactly to observed data rather than using models based on assumptions, and increasing complexity does not present a great problem, simply more work. The disadvantage is that we are no longer able to use mathematical analysis and instead we have to resort to simulations and to rely on statistical results from our models.

When making our choice of model, the first important consideration is whether we are dealing with a *discrete* or a *continuous* random variable. A *discrete* random variable is one that can take any value from a set of distinct values but nothing in between. For example the number of children in a family has to be a non-negative integer. A *continuous* random variable, on the other hand, can take any value within a certain range, e.g. a child's height. In practice the distinction between the two types is often blurred because continuous variables will be measured to the nearest unit on some scale of measurement, e.g. height to the nearest centimetre. One of the modelling decisions that we may have to make is whether to represent a particular variable as continuous or discrete.

Theoretical models for *discrete* random variables are specified by a *probability function* $p(x) = \text{Prob}(X = x)$, which is the probability that the random variable X takes the value x. The simplest model is the *discrete uniform distribution* where each value has the same probability. This is what we would probably use if X represented the score on an ordinary six-sided die as used in games of chance.

A very useful discrete model is the *binomial distribution.* If there are n independent items, each of which has a constant probability p of being a certain type, then the number X of items of that type is a random variable whose probability function is

$$p(x) = \binom{n}{x} p^x(1-p)^{n-x}, \qquad x = 0, 1, 2, \ldots, n.$$

Suppose that events occur at random such that in a small time interval Δt the probability that an event occurs is $\lambda\,\Delta t$ and the probability that more than one event occurs is zero. If X is the number of events occurring in one unit of time, X is a random variable whose probability function is

$$p(x) = \frac{\lambda^x \exp(-\lambda)}{x!}, \qquad x = 0, 1, 2, \ldots .$$

This is the *Poisson distribution* and the sequence of events is described as a Poisson process.

Data on a discrete random variable can be put into a relative frequency table as we did for the first example in this section. When the number of observations is large, we expect the relative frequencies to correspond closely to the values of the probability function.

Theoretical models for a *continuous* random variable are specified by a *probability density function* (pdf) $f(x)$ where $f(x) \geqslant 0$ for all x and

$\int_{-\infty}^{\infty} f(x)\,dx = 1$. The probability that the random variable takes a value between x_1 and x_2 is given by $\int_{x_1}^{x_2} f(x)\,dx$.

The simplest example is the continuous uniform distribution on the interval $[a, b]$ (Fig. 7.1). For this,

$$f(x) = \begin{cases} \dfrac{1}{b-a}, & a \leqslant x \leqslant b, \\ 0, & \text{otherwise.} \end{cases}$$

For example, if θ is the angle that a spinning pointer makes with a fixed direction when it finally comes to rest, then (unless the pointer has a tendency to stick in a particular direction) θ is a random variable and the appropriate model is the continuous uniform distribution on $[0°, 360°]$.

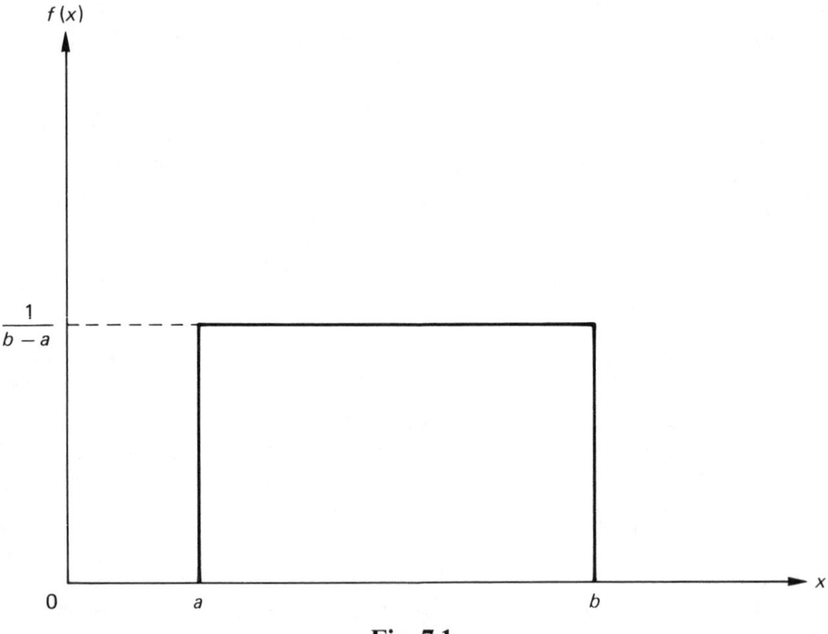

Fig. 7.1

In the Poisson process described earlier, the time gap between consecutive events is obviously a continuous random variable. Its pdf is $f(x) = \lambda \exp(-\lambda x)$, $x \geqslant 0$, and it is known as the exponential distribution (Fig. 7.2).

Note that small time gaps are quite common (high probability) while long time gaps are rare. The mean time gap is $1/\lambda$.

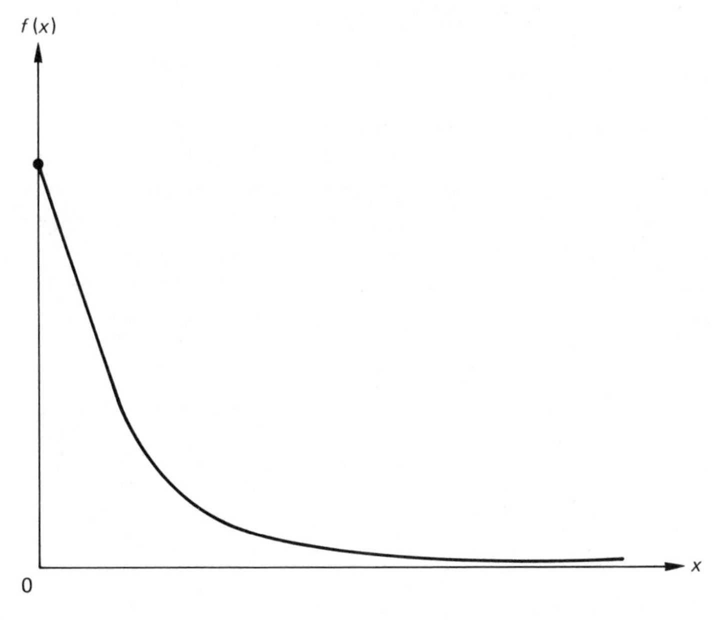

Fig. 7.2

Another very useful and popular model for continuous random variables is the Normal distribution (Fig. 7.3). The pdf is

$$f(x)=\frac{1}{\sigma\sqrt{2\pi}}\exp\left[-\left(\frac{x-\mu}{\sigma}\right)^2\right].$$

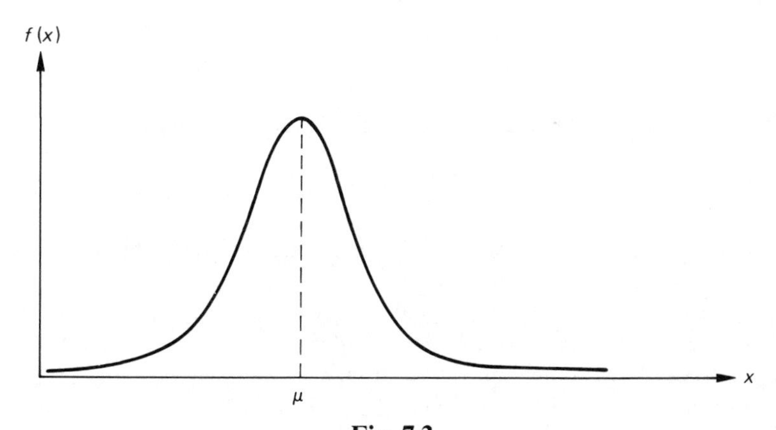

Fig. 7.3

There are two parameters μ and σ; μ is the mean and determines the central position while σ is the standard deviation and determines the 'width' of the curve.

Data on a continuous random variable can be put into a relative frequency table just as in the case of a discrete random variable but the difference is that in the case of a discrete random variable each relative frequency corresponds to a particular value of the random variable. For a continuous random variable, each relative frequency corresponds to a *range* of values of the random variable. For example a student finds that his journey to college can take up to 20 min. He records the time t on a number of occasions and tabulates the data as follows.

t/min	0–10	10–15	15–20
Relative frequency	0.2	0.5	0.3

This is an example of a frequency table (or, when plotted graphically, a histogram) for a continuous random variable.

7.3 GENERATING RANDOM NUMBERS

The toss of a fair coin produces a random outcome which can be used as an observation of the discrete random variable X whose probability function is the following.

x	0	1
$p(x)$	0.5	0.5

If we needed to *simulate* a value or a set of values of this random variable, we could do it by tossing the coin and recording the results. This is rather limiting, however, and for random variables with other distributions we use computer-generated pseudo-random numbers. These are produced in a sequence based on a formula. The numbers obtained do not have any obvious pattern and, when suitably scaled, it is found that their distribution corresponds very closely to the uniform distribution on the interval [0, 1].

An example of such a formula is

$$X_{n+1} = 97X_n + 3 \pmod{1000},$$

$$R_{n+1} = \frac{X_{n+1}}{1000}.$$

An arbitrary starting value ('seed') is taken—for example $X_0 = 71$; then, to get X_1, we substitute for X_0 in the formula, which gives 6890. Next we count any multiples of 1000 as being equivalent to zero, which gives 890. We then divide by 1000 to give $R_1 = 0.890$. Substituting $X_1 = 890$ gives $R_2 = 0.333\ldots$ and so on for as long as we like. (Eventually we shall get back to 71 and repeat the cycle but only after a very long time.)

In many high-level programming languages and on pocket calculators a standard function usually represented by some name such as RND produces a pseudo-random number of this kind. From now on, we shall use RND to stand for a generated random value from the uniform continuous distribution on $[0, 1]$.

We can build other random variables from this. For example $X = a + (b - a)\,\mathrm{RND}$ gives a random variable whose distribution is the continuous uniform distribution on the interval $[a, b]$.

If we want to produce values for the exponential distribution with parameter λ, we use $X = -(1/\lambda)\ln(\mathrm{RND})$ or, more conveniently, $X = -m\ln(\mathrm{RND})$, where m is the mean.

Values for the Normal distribution with a mean of 0 and a standard deviation of 1 can be produced by taking two RND values and then using the formulae

$$X_1 = [-2\ln(\mathrm{RND}_1)]^{1/2}\cos(2\pi\,\mathrm{RND}_2) \text{ and}$$

$$X_2 = [-2\ln(\mathrm{RND}_1)]^{1/2}\sin(2\pi\,\mathrm{RND}_2)$$

to give two values of X. If values from the Normal (μ, σ^2) distribution are required, then use $X = \sigma X_1 + \mu$ or $\sigma X_2 + \mu$.

To obtain values for *discrete* random variables, divide the $[0, 1]$ range into appropriate parts. For example, suppose a discrete random variable has the following probability function.

x	0	1	2
$p(x)$	0.3	0.3	0.4

Take an RND value; then, if $0 < \mathrm{RND} < 0.3$, this means $X = 0$, if $0.3 < \mathrm{RND} < 0.6$, this means $X = 1$, and, if $\mathrm{RND} > 0.6$, this means $X = 2$.

RND	0	0.2	0.4	0.6	0.8	1.0
X	← $X=0$ →		← $X=1$ →	← $X=2$ →		

This ensures that the values 0, 1 and 2 come up statistically with the right frequencies.

We use the same idea for a *continuous* random variable tabulated into classes except that we take the generated value of the random variable to be the class midpoint.

X	0–10	10–15	15–20
Relative frequency	0.2	0.5	0.3

An RND value of 0.36 would be translated into $X = 12.5$. A refinement would be to use linear interpolation rather than just to take the midpoint, i.e.

$$\frac{X-10}{5} = \frac{0.36-0.2}{0.5},$$

giving

$$X = 11.6.$$

To simulate a theoretical distribution for a continuous random variable with known pdf, there are two main methods available.

1 *The inverse cumulative distribution function method* If the pdf of a random variable is $f(x)$, the cumulative distribution function is $F(x) = \int_{-\infty}^{x} f(t)\,dt$ and if regarded as a random variable, F is uniformly distributed on $[0, 1]$. A value of RND is taken from the uniform $[0, 1]$ distribution; then the equation $\text{RND} = F(x)$ is solved to find the corresponding value of x $(= F^{-1}(\text{RND}))$. For example, suppose that

$$f(x) = \begin{cases} 0.5 \sin x, & 0 < x < \pi, \\ 0, & \text{otherwise.} \end{cases}$$

The graph is shown in Fig. 7.4.

The cumulative distribution function is

$$\begin{aligned} F(x) &= \int_0^x 0.5 \sin t\,dt \\ &= [-0.5 \cos t]_0^x \\ &= 0.5(1 - \cos x) \end{aligned}$$

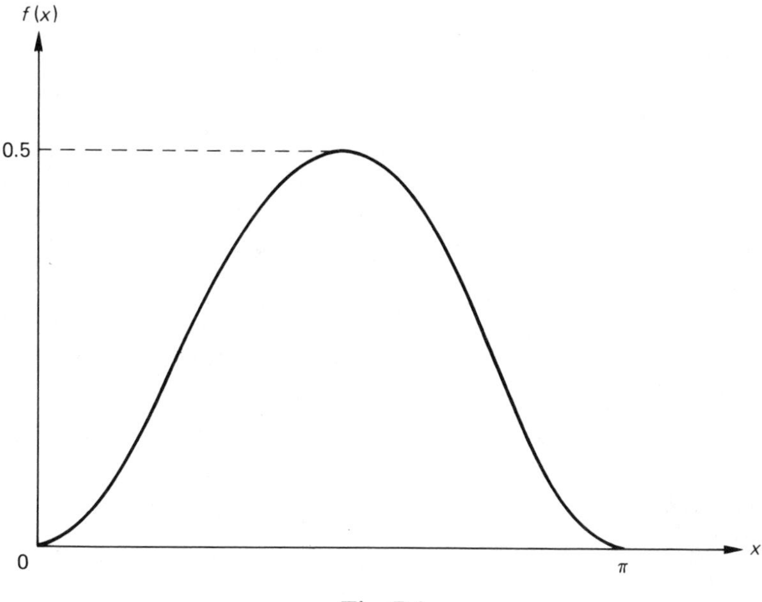

Fig. 7.4

Inverting $\text{RND} = 0.5(1 - \cos X)$ gives

$$X = \arccos(1 - 2\,\text{RND}).$$

This is the formula we need for generating X values for this distribution.

2 *The rejection method* For this, we need two RND values to generate one X value. Suppose that $f(x) = 0$ except on the interval $[a, b]$ and that the maximum value of $f(x)$ is c (Fig. 7.5). To generate an X value, we go through the following steps.

(a) Generate RND_1 and RND_2 from the uniform $[0, 1]$ distribution.
(b) Use RND_1 to calculate $x = a + (b - a)\,\text{RND}_1$.
(c) Calculate $f(x)$.
(d) Use RND_2 to calculate $y = c\,\text{RND}_2$.
(e) If $y < f(x)$, then accept x; otherwise reject x and return to step (a).

For the above example, we would use $a = 0$, $b = \pi$ and $c = 0.5$.

You can see that the RND function provides us with the raw material that we need to generate sample values of random variables. This is called 'simulation' and it breathes life into our probabilistic models. Note that in this chapter we are using the term 'simulation' as short-hand for 'stochastic simulation' (see section 3.2 for a wider discussion on simulation).

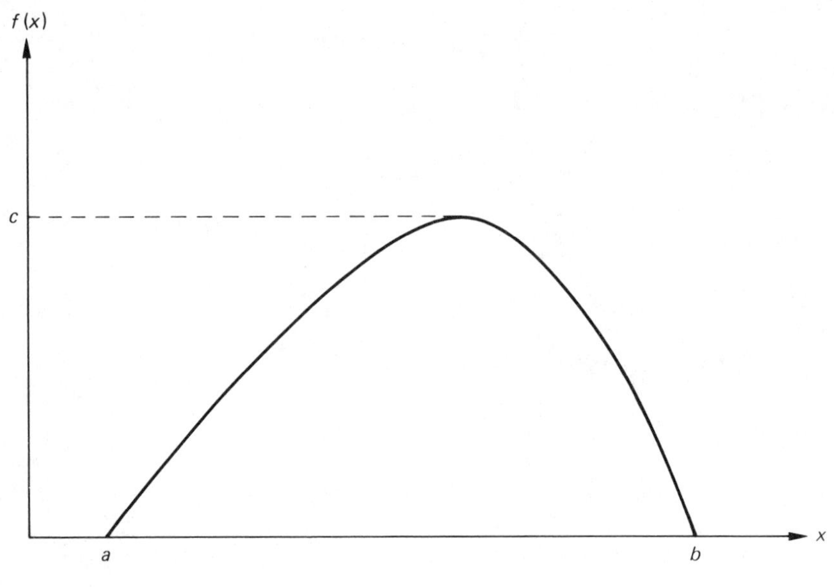

Fig. 7.5

7.4 SIMULATIONS

What we mean here by a simulation is a probabilistic model set in motion by a stream of random numbers. As a simple example, consider the following problem.

Example 7.4.1

A train leaves station A at a time which is about 1.00 pm but distributed as follows.

Time of departure	1.00 pm	1.05 pm	1.10 pm
Relative frequency	0.7	0.2	0.1

The train's next stop is at B where you are hoping to catch it. The train's journey time from A to B is variable with a mean value of 30 min and a standard deviation of 2 min. Your time of arrival at B is distributed as follows.

Time	1.28 pm	1.30 pm	1.32 pm	1.34 pm
Relative frequency	0.3	0.4	0.2	0.1

What is the probability that you catch the train? An equivalent way of posing the question is on what percentage of occasions under these identical conditions will you catch the train? It is this second (relative frequency) interpretation which we use to obtain an answer by simulation. We need random numbers; so we shall use those given by the formula in section 7.3, which are 0.890, 0.333, 0.304, 0.491, 0.630, ... What we need to simulate are the following.

1 The departure time t_1 of the train.
2 The journey time t_2 of the train.
3 Your arrival time t_3 at the station.

We are interested in whether $t_3 < t_1 + t_2$.

To run the simulation, we simply need to produce values for t_1, t_2 and t_3 and to test the condition. It is not obvious how to get a value for t_2. We do not know the distribution; so we shall have to assume a model. Let us take the normal distribution with a mean of 30 min and a standard deviation of 2 min. Note that we are treating t_1 and t_3 as discrete and t_2 as continuous. This is only because of the way that our data were presented.

Using a minute as the time unit and counting from $t = 0$ at 1.00 pm our model is made up as follows:

$$\left.\begin{array}{rl} 0 < \text{RND} < 0.7, & t_1 = 0, \\ 0.7 < \text{RND} < 0.9, & t_1 = 5, \\ 0.9 < \text{RND} < 1.0, & t_1 = 10; \end{array}\right\}$$

$$\left.\begin{array}{rl} 0 < \text{RND} < 0.3, & t_3 = 28, \\ 0.3 < \text{RND} < 0.7, & t_3 = 30, \\ 0.7 < \text{RND} < 0.9, & t_3 = 32, \\ 0.9 < \text{RND} < 1.0, & t_3 = 34; \end{array}\right\}$$

$$t_2 = 2X + 30,$$

where $X = [-2\ln(\text{RND}_1)]^{1/2} \cos(2\pi\,\text{RND}_2)$. The calculation gives the following table.

RND	t_1	t_2	t_3
0.890	5	—	—
0.333 } 0.304 }	—	29	—
0.491	—	—	30

So, on this occasion, you arrive 4 min before the train comes in. Note that we must not use the *same* RND value for more than one purpose. This would lead to false correlations between variables in our model.

The conclusion from this single run of the model is that you caught the train, but obviously this does not answer the original question. We need to do several runs and to record the percentage of runs on which we catch the train. The answer that we get can only be regarded as an approximation. To find the exact answer by this method, we would have to carry out a theoretically infinite number of runs. This is a feature of all simulation models. We must carry out many runs and calculate averages from our results. Clearly, it will be a great help if we implement our model in the form of a computer program.

The above example is rather trivial and far from typical of the kind of model for which simulation is usually used. In practice, we normally refer to the set of circumstances which we are trying to model as a 'system'. We shall take a simple model of a hairdresser's shop as an example of a system.

Example 7.4.2

Customers arrive at a hairdresser's shop at random times. There are two assistants, A and B. 60% of customers require a cut, which takes 5 min; the other 40% require a shampoo and cut, which takes 8 min.

Generally speaking, the basic elements needed for carrying out any simulation are as follows.

(a) A set of *state variables* whose values completely describe the system at any time.
(b) A procedure for calculating new values of the state variables at time $t + 1$ from those at time t.

In our example there are three state variables.

1 The number of customers waiting (discrete non-negative integer).
2 Whether or not A is busy (yes or no).
3 Whether or not B is busy (yes or no).

A *run* of the simulation consists of a series of calculations starting with the values of the state variables at $t = 0$ and ending at some time $t = \text{END}$. An *event* is a point in time when a state variable changes its value. In our example the events are as follows.

(i) An arrival.
(ii) Start of service by A.
(iii) End of service by A.
(iv) Start of service by B.
(v) End of service by B.

An *entity* is a discrete item which either is a permanent part of the system or enters and leaves, e.g. a customer, a lorry or an order. Here the entities are the customers and the two assistants.

There are two types of procedure for studying a simulation model.

1 *Time slicing* where we examine the state variables and the whereabouts of the entities at time slices (usually equally spaced). At each time slice the state variables may or may not have changed.
2 *Event sequencing* where we examine the system at each event, regardless of the time between events.

These two approaches are sometimes referred to as 'time-driven' and 'event-driven' models. Generally, we use time-driven models for continuous deterministic systems and event-driven models for probabilistic discrete systems, but this is not always the case and for this particular example we shall use both types.

For the *time-slicing* model, we must decide how large to make the time slices and for simplicity we shall take 1 min. The description of the problem did not contain any information about the rate of arrival of customers. Suppose that in any minute the probability that a customer arrives is 0.5. There are actually two kinds of customer, depending on whether they have come for a cut or for a shampoo and cut. We can make up a crude and simple model by taking the average service time $0.6 \times 5 + 0.4 \times 8 = 6.2$ min and taking this to apply to all customers.

A coin will do as a random-number generator, where T is tails and H heads. Suppose that a sequence of tosses gives THTTHTTTHHT.... Taking T to indicate an arrival and taking the initial state to be no customers, we run through the first 10 min as follows.

Time/min	Arrival?	A busy?	B busy?	Queue
0	No	No	No	0
1	Yes	Yes	No	0
2	No	Yes	No	0
3	Yes	Yes	Yes	0
4	Yes	Yes	Yes	1
5	No	Yes	Yes	1
6	Yes	Yes	Yes	2
7	Yes	Yes	Yes	3
8	Yes	Yes	Yes	3
9	No	Yes	Yes	3
10	No	Yes	Yes	3

At this point, we shall stop to ask what we hope to get out of it. The usual things of interest are the average queue length, the maximum queue length, the average waiting time of customers, and the percentage busy times of the two assistants. We shall look at how to calculate these results in section 7.5. The point of creating the model is, of course, not just to try to mimic the real system but to obtain answers to *questions*. These will usually be questions about the effects of making changes such as employing an extra assistant, or altering the service time.

In our simple model, we have made a number of assumptions as well as gross simplifications.

1 We assumed (for our own convenience) that the probability of an arrival in any minute is 0.5. If we had data indicating that a more accurate figure would be 0.3, we could no longer use the coin. We would use the RND function, taking RND > 0.3 to mean that a customer arrives.
2 We implicitly assumed that more than one customer would never arrive in the same minute. If this had been observed to happen, it would mean that our choice of 1 min as the time slice was not good; a smaller time would be better.
3 We assumed that, if both assistants were free, a customer would make a random choice between them. If necessary, we could build in a preference into our model.
4 We assumed a 'queue discipline' of 'first in, first out' (FIFO). Some customers might have prior bookings and would get priority.
5 In practice a customer might, on arriving and finding a long queue, decide not to stay. A customer who had been waiting a long time might also get tired of waiting and leave without being served. One of the assistants might take a short break. We could include all these possibilities in the model.

An *event-sequencing* model requires more work but enables us to get closer to the real situation. We use the term *activity* to refer to the time between two events. In our problem the activities are as follows.

(a) The queueing time.
(b) The service time by A.
(c) The service time by B.

The *duration* of an activity may be constant or random. If random, then we need a distribution to model it. We also need to specify how many entities can use the activity in parallel. In our example there can be many customers waiting simultaneously but an assistant can only serve one at a time. Many systems can be regarded as consisting of entities passing through a number of activities. The paths taken by the entities can be shown in a *flow diagram*. If more than one entity can be engaged on an activity in parallel, we must show this in our flow diagram, and whether or not there is an upper limit.

An *attribute* is a property of an individual entity, e.g. the size of an order, or whether a lorry is full or empty. The attribute could be constant or could change during the simulation and there could be a number of attributes attached to each entity. In our example each customer has only one attribute—the type of service required.

It may be useful to think of a *queue* as a special kind of activity. The most obvious feature is that there will often be many entities engaged in this activity at the same time. There also has to be a *queue discipline* such as 'last in, first out' (LIFO) or FIFO and, when there is more than one queue leading to the same activity, we need rules to state which queue has priority. We may also need priority rules within a single queue. These may be based on the attributes of the entities, e.g. 'empty lorries have priority'. Queues are different from other activities in that the length of time that an entity stays in a queue depends on factors such as the number of other entities in the queue, priorities and rules. We cannot just sample the queuing time from a distribution.

Another special activity required is the *generating* activity for generating arrival events. This activity represents the inter-arrival time and we often use the exponential distribution as a model.

To carry out the simulation, we generally have the following options.

1 Work it through by hand.
2 Use a specially written computer program.
3 Use a computer package or simulation language.

Option 1 is obviously the most tedious and least efficient but it has the merit of helping us to sort out our ideas clearly. It is often a good idea at least to start a hand simulation until we are confident that we have modelled all the necessary details correctly. We can then turn to option 2 or option 3.

The standard procedure for carrying out a hand simulation involves the following steps.

1 Make a list of the entities (and attributes if any), state variables, activities and queues. For each activity, write down starting conditions (if any), method of determining duration, and the maximum allowed number of entities in parallel. For each queue, state the queue discipline and any other operating rules. Write down the starting conditions and the end time (or ending condition) for the simulation.

For the hairdresser's example, let us assume that the inter-arrival time has the exponential distribution with a mean of 3 min. We can then simulate the time between arrivals by $-3\ln(\text{RND})$ (min). Let us work through the simulation as far as the arrival of the tenth customer. We list the complete problem as follows.

Entities	Customers (attribute: $P(\text{cut}) = 0.6$; $P(\text{wash and cut}) = 0.4$). Assistant A. Assistant B.
Activities	Inter-arrival (exponential; mean = 3 min). Queue (unlimited; FIFO; no priorities). Service by A (cut, 5 min; wash and cut, 8 min). Service by B (cut, 5 min; wash and cut, 8 min).
Events	Arrival. Start of service by A. End of service by A. Start of service by B. End of service by B.
Starting conditions	Queue = 0; A idle; B idle.
End condition	Arrival of tenth customer.

2 Draw a flow diagram (Fig. 7.6) showing the events and activities. This shows the paths that we have to trace out as we follow through the simulation and it can be useful in constructing a computer program. Remember that events are points in time, and activities are things that *take* time.

3 Draw up a table ('trace table') which will enable you to step through the simulation from event to event. This should include the time since the start of the simulation ('CLOCK'), the next event and the state variables on each line.

4 Produce as many RND values as you need. You may use these as you wish, provided that you do not use the same value twice. Either produce

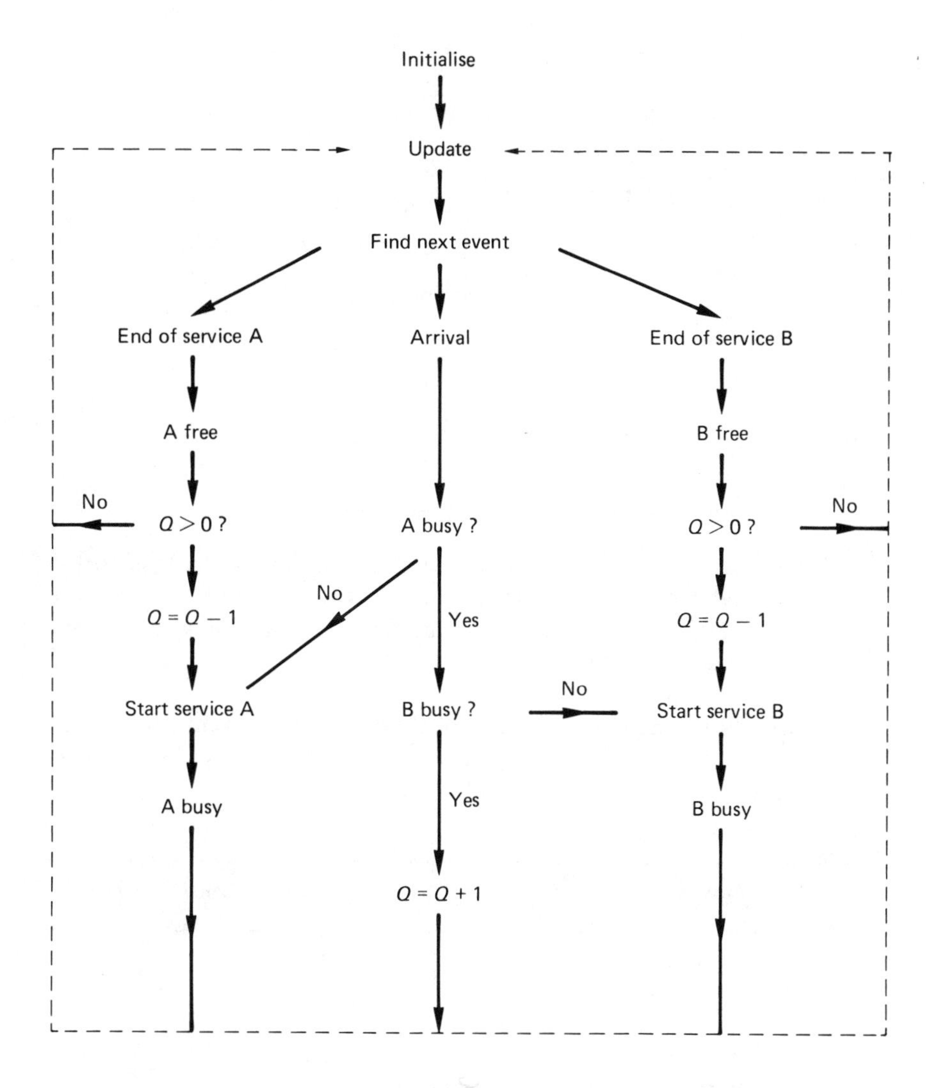

Fig. 7.6

them as and when they are needed or produce them at the beginning of the simulation and convert them into the required information. For our hairdresser's shop we need a total of 19 RND values, nine of them for the inter-arrival times and one for each customer to find out whether they have come for a cut ($\text{RND} < 0.6$) or a wash and cut ($\text{RND} > 0.6$).

We have used the random-number sequence from section 7.3 and converted the first nine (from 0.071 to 0.511) into arrival times by the formula $-3\ln(\text{RND})$. The next 10 (from 0.570 to 0.853) have been used to decide the type of service required for each customer.

RND	−3 ln(RND)	Arrival	CLOCK	RND	Cut or wash and cut?
0.071	7.935	1	0	0.570	Cut
0.890	0.350	2	7.935	0.293	Cut
0.333	3.299	3	8.285	0.424	Cut
0.304	3.572	4	11.584	0.131	Cut
0.491	2.134	5	15.156	0.710	Wash and cut
0.630	1.386	6	17.290	0.873	Wash and cut
0.113	6.541	7	18.676	0.684	Wash and cut
0.964	0.110	8	25.217	0.351	Cut
0.511	2.014	9	25.327	0.050	Cut
		10	27.341	0.853	Wash and cut

Note that we have started the CLOCK at the time of the first arrival and accumulated the inter-arrival times to get the CLOCK times for subsequent arrivals. We have kept accuracy to three decimal places, which probably was not necessary.

5 Fill in the trace table, taking care to identify the next event correctly at each step. We have used the abbreviation EOS for 'end of service' and Arr. 2 for 'Arrival number 2', etc.

CLOCK	Next event	Q	Start A	End A	Start B	End B
0	EOS A at 5.000	0	0	5.000	—	—
5.000	Arr. 2 at 7.935	0	—	—	—	—
7.935	Arr. 3 at 8.285	0	—	—	7.935	12.935
8.285	Arr. 4 at 11.584	0	8.285	13.285	7.935	12.935
11.584	EOS B at 12.935	1	8.285	13.285	7.935	12.935
12.935	EOS A at 13.285	0	8.285	13.285	12.935	17.935
13.285	Arr. 5 at 15.156	0	—	—	12.935	17.935
15.156	Arr. 6 at 17.290	0	15.156	23.156	12.935	17.935
17.290	EOS B at 17.935	1	15.156	23.156	12.935	17.935
17.935	Arr. 7 at 18.676	0	15.156	23.156	17.935	25.935
18.676	EOS A at 23.156	1	15.156	23.156	17.935	25.935
23.156	Arr. 8 at 25.217	0	23.156	31.156	17.935	25.935
25.217	Arr. 9 at 25.327	1	23.156	31.156	17.935	25.935
25.327	EOS B at 25.935	2	23.156	31.156	17.935	25.935
25.935	Arr. 10 at 27.341	1	23.156	31.156	25.935	30.935
27.341	EOS B at 30.935	2	23.156	31.156	25.935	30.935

Note that we have assumed that as soon as an assistant has finished serving one customer he or she instantly takes another one (if there is one) from the queue. Note also that the queue does not include customers being served. Many mathematical modellers do include the customer being served as being still in the queue. Watch out for this and make sure that you have explained your method clearly.

It seems rather unrealistic to assume constant service times of 5 and 8 min. If we wanted to allow variable service times, how would we do it? The choices are either to assume some models for the distributions, such as a uniform distribution of between 4.5 and 5.5 min for a hair cut or, if data are available, to take sample values using the relative frequencies as we did in section 7.3. We could obviously assume different distributions for the two assistants instead of assuming that they are identical as we have done.

EXERCISES 7.4

1 Obtain an answer for Example 7.4.1 (the probability of catching the train).

2 Run the hairdresser's shop model using the following information.

(a) When the shop opens, there are no customers waiting.
(b) The inter-arrival time of customers has the exponential distribution with a mean of 3 min.
(c) 60% of customers require a cut and 40% require a shampoo and cut.
(d) Assistant A's service time is exponentially distributed with a mean of 5 min for cutting and a mean of 8 min for washing and cutting.
(e) Assistant B's service time is exponentially distributed with a mean of 4.5 min for cutting and a mean of 7 min for washing and cutting.
(f) Customers finding a queue of six people or more when they arrive will not stay.
(g) Both assistants will take a 1 min rest after serving five customers without a break.

Run the model until 30 customers have passed through the system. Calculate the average waiting time, the average queue length and the number of customers who turned away. Also calculate the percentage busy time for both assistants. How would these figures change if

(i) assistant A's mean service times could be reduced to those of B?
(ii) the mean inter-arrival time became 2 min or less?
(iii) a third assistant with the same mean service times as A was employed?

7.5 USING SIMULATION MODELS

Having developed a model which seems to simulate a system with reasonable success, do not think the major part of the work is over! A single run of the model tells you nothing more than that the model appears to be behaving as it should. To make use of the model, you need to make several runs and to analyse the results. The kind of analysis which may be needed depends on the objectives for the model and these should be stated clearly at the start. (In our hairdresser's example, our only objective was to go through the mechanics of developing a simulation model.)

In practice the objectives for a simulation model can be varied but very often they include some of the following.

1 *Collecting statistics on the long-term behaviour of a system* If we are trying to simulate the operation of a port, for example, we may well be interested in extremes of congestion which may occur from time to time and which will be revealed in a long-term study of the port's operation. We can, of course, run through a very long period of simulated time in a few minutes of real computing time.
2 *Comparing alternative arrangements of the system* This is where a good simulation model pays dividends. Making alterations to a complicated system can be very costly and time consuming in the real world. In the model world, we can make alterations quickly and easily. Examination of the results will help in deciding whether or not to go through with implementing the changes in the real world.
3 *Investigating the effects of changing parameters* We can use a model to find what happens if for example the mean time between arrivals decreases. Will the system be able to cope? What happens if the service times are increased?
4 *Investigating the effects of altering the modelling assumptions* We may have used a rather simple model and are wondering whether it will make a significant difference if we improve it slightly.
5 *Finding optimal operating conditions for the system* There may be a number of different possible arrangements of services and queues within a system and we want to find out the 'best' arrangement. We shall obviously need some kind of *performance measure* which we can use to identify the best arrangement.

A constant problem with simulation models is the choice of *initial conditions*. What happens in the model depends, sometimes for a surprisingly long time, on the initial state. If we start from a rather 'artificial' state such as having no queues and no customers in a system, we need to run the model for some time before we get into a more typical 'busy' state. In many systems, this 'typical' behaviour, once we have reached it, carries on indefinitely and is known as the *steady state*. The build-up period before reaching the steady state is referred to as the *transient* stage.

Before running your model, consider whether you want to simulate the steady state or transient behaviour of the system. If it is the transient stage, then you must think carefully about the starting conditions. For the steady state, unless you know what a typical state of the system looks like, your only option is to put in an arbitrary set of starting values and to run the model for long enough for the effects of your starting values to be wiped out. In this case, when you come to analyse the results, you will of course ignore the data covering the transient stage. How many events you will have to generate to reach the steady state is impossible to say; you have to experiment.

To carry out a simulation study, we make several runs of the model and collect statistics from each run. In particular, we want statistics on the *performance measures* mentioned earlier. Exactly what these are depends on the problem and on the point of view that we wish to adopt. From the point of view of customers the important performance measures are the following.

(a) The average waiting time.
(b) The average queue length.
(c) The maximum queue length.

From the point of view of the system, it may be more important to look at the percentage idle time of servers and customer turn-over rate.

Note that for different runs you should use different random-number streams except when comparing two systems, when it would be useful to have exactly the same arrival pattern, for example. Note also that we talk about 'runs' without specifying how long each run is. Again this is something that can be decided after experimenting with your model. You can obviously choose either of two ways of ending a run: after a specified CLOCK time has elapsed or when a certain condition is satisfied. In a model involving simple arrivals, for example, you could end the run after a specified number of arrivals.

Let us now take a more detailed look at how to collect the relevant statistics, using the hairdresser's example again. Suppose that our original objective had been to calculate the following.

(i) The average waiting time of customers.
(ii) The average queue length.
(iii) The maximum queue length.
(iv) The percentage busy times of the two assistants.

Let us run the simulation from CLOCK = 0 to CLOCK = END. Note

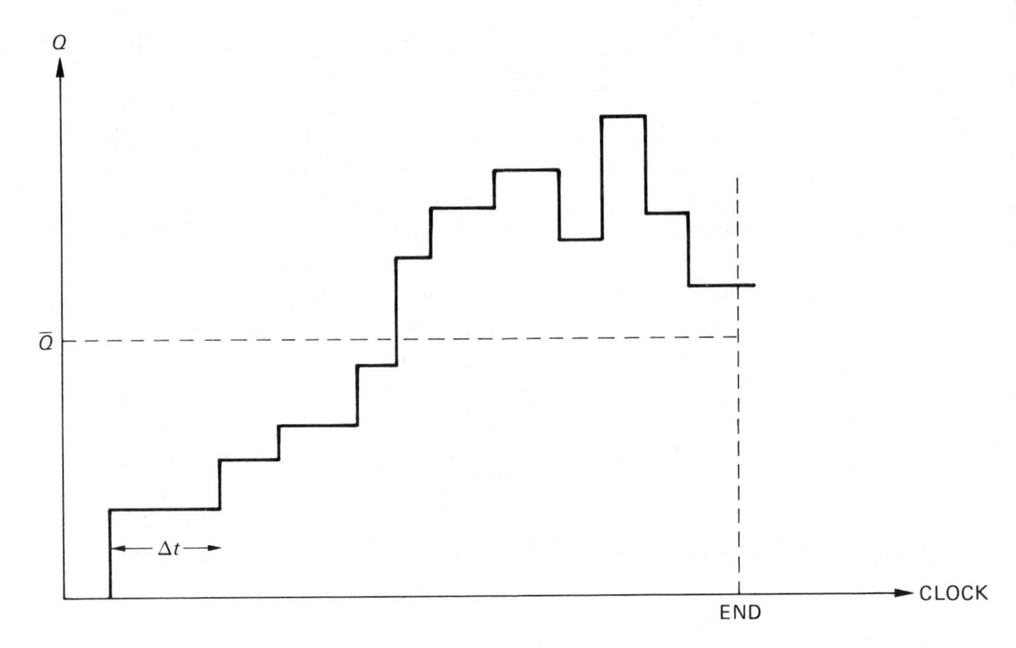

Fig. 7.7

that there are two quite different sorts of averages involved here: an average over *time* and an average *per customer*. Let Q be the number of customers queueing at any time. The graph of Q against CLOCK time might look like Fig. 7.7.

The average value of Q over time is $\bar{Q}$, where $\bar{Q} \times \text{END} =$ the total area under the graph. Let Δt represent an interval of time over which Q remains constant (where Δt itself is variable of course). As we run through the simulation, we accumulate the sum $\sum(Q\,\Delta t)$. Let N denote the total number of arrivals during the run. The two averages that we need are given by

$$\text{average queue length } \bar{Q} = \frac{\sum(Q\,\Delta t)}{\text{END}},$$

$$\text{average waiting time } \bar{W} = \frac{\sum(Q\,\Delta t)}{N}.$$

Here is part of the trace table for the hairdresser's model showing the accumulated queueing time $\sum(Q\,\Delta t)$. Note that it only shows the times at which Q changes.

CLOCK	Q	t	$Q\,\Delta t$	$\sum(Q\,\Delta t)$
0	0	0	0	0
11.584	1	0	0	0
12.935	0	1.351	1.351	1.351
17.290	1	4.355	0	1.351
17.935	0	0.645	0.645	1.996
18.676	1	0.741	0	1.996
23.156	0	4.480	4.480	6.476
25.217	1	2.061	0	6.476
25.327	2	0.110	0.110	6.586
25.935	1	0.608	1.216	7.802
27.341	2	1.406	1.406	9.208

The total elapsed time END is 27.341. The number N of arrivals in this time is 10. The maximum queue length Q_{max} was 2. The accumulated queueing time is $\sum(Q\,\Delta t) = 9.208$. The mean queue length $\bar{Q}$ is $9.208/27.341 \approx 0.34$. The mean waiting time $\bar{W}$ is $9.208/10 \approx 0.92$ min.

We ought to mention that we terminated the simulation with two customers being served—one in the queue and one newly arrived.

The total time for which server A was busy was $(5-0)+(13.285-8.285)+(23.156-15.156)+(27.341-23.156)$ min $= 22.185$ min. Assistant A was therefore busy for $(22.185/27.341) \times 100\% \approx 81\%$ of the time. Assistant B was busy for a total of 19.406 min or about 71% of the time.

Note that it may not be sufficient just to calculate averages because these give no indication of *variability*. For each run, we can give more statistical detail by means of histograms showing the *distribution* of waiting times, for example (i.e. how many customers waited how long). We cannot derive histograms from the $\sum(Q\,\Delta t)$ value, however, and it is necessary to record the arrival and leaving times of each customer and to store them in an array.

To give a measure of variability, we can calculate standard deviations. If we have carried out a number of runs, we can use the variation in the $\bar{Q}$ and $\bar{W}$ values for individual runs to compute *interval estimates* for the true values of $\bar{Q}$ and $\bar{W}$ besides giving us an idea of how much reliance we can put on the results of a single run.

EXERCISES 7.5

1 The manager of a bank is considering changing over to a new single-queue system but is not sure whether it will be better for customers than the existing system. In the present system there are five service points and, when customers enter the bank, they can choose to queue at any of the

five points. If a customer at the back of any queue notices that another cashier has become free, he will move over for service. The average time between the arrivals of customers in the busy period is M min and the average time taken to serve a customer is 2.5 min.

Devise a simulation model which will enable you to compare the two systems and help the bank manager to make his decision. How does the conclusion depend on the value of M? Can you include a model for the variation in the value of M during the working day?

2 A small supermarket has four cash tills and the time taken to serve a customer at any of the tills is proportional to the number of items in his or her trolley (roughly 1 s per item). 20% of customers pay by cheque or credit card which takes 1.5 min; paying with cash takes only 0.5 min. It is proposed to make one cash till a quick-service till for customers with eight items or less. Two of the remaining tills are to be designated 'cash only'.

You are asked to develop a simulation model which will enable you to compare the operation of the proposed system with that of the present system. Assume that the mean time between arrivals of customers is 0.5 min and that the number of items bought by customers is represented by the following frequency table.

Number of items	<8	9–19	20–29	30–39	40–49	>50
Relative frequency	0.12	0.10	0.18	0.28	0.20	0.12

3 An office has three telephone lines for incoming calls, which effectively means that a maximum of three customers can be dealt with at any one time. Customers ring at random times uniformly distributed between 9.00 am and 5.00 pm. Each call lasts for a random length of time but the average is 6 min.

The manager is worried about the possible number of customers who are unable to get through because all three lines are engaged. A certain percentage of these might try again later but some may not bother. The average number of calls received in a day is 70.

You are asked to help to put the manager's mind at rest by developing a model to simulate the telephone calls. Use your model to estimate the following.

(a) The percentage of time for which zero, one, two and three lines are engaged.
(b) The percentage of customers who are put off.

How would you adapt the model if the office had an extra line installed? What extra information would you need to improve the model?

7.6 PACKAGES AND SIMULATION LANGUAGES

As you will have gathered from the previous sections, simulation is essentially an experimental technique in that models are set up and 'run' to see what happens. It makes things much easier if the 'running' happens inside a computer and it is not surprising that there is a lot of computer software available for constructing and running stochastic simulation models.

The software can be divided into simulation *packages* and simulation *languages*. The distinction is that a *package* will have some form of user interface which will allow you to define your model either by giving responses to questions asked in ordinary English or by making selections from prepared menus. A simulation *language*, on the other hand, is a specially written high-level language in which you will write your model as a sequence of statements.

Both types of software offer the facilities normally required in discrete simulation such as the generation of pseudo-random numbers for any desired statistical distribution, the collection and summary of data from runs and the statistical analysis and presentation of results. The disadvantage of a simulation language is that you have a new language to learn, which takes up some time. The disadvantage of a simulation package is that it tends to be very convenient for simulating a particular class of problem (for which it has been designed) but very difficult to use on a problem of a completely different kind.

A wide variety of packages and simulation languages are available, differing in their power, flexibility and ease of use. At present the most widely used languages are SIMSCRIPT, GASP, GPSS and SIMULA. You can of course write your own simulation programs in whatever language you choose but the fact that all stochastic simulations involve common elements such as generating random numbers, finding the next event and updating queues means that you will, to some extent, be constantly 'reinventing the wheel'.

7.7 SUMMARY

In this chapter, we have taken a superficial look at a very important type of modelling, namely modelling involving random numbers. This is a very large area and what we have described is merely a brief introduction. There are two different approaches which can be taken in modelling of this type.

1 The analytic approach is to use the theory and functions of mathematical statistics.
2 The simulation approach is to create a model having the required structure and then to use a random-number generator to activate the model. This is often referred to as 'discrete event simulation' or 'stochastic simulation' and is commonly 'event-driven' rather than 'time-driven'.

The type 1 approach becomes very difficult, once a model acquires a reasonable amount of complexity. The type 2 approach has the disadvantage that several runs are necessary and the results obtained are statistical in nature.

Event-driven simulations can be done by hand but become rather tedious and it is usual to turn to computer software for help. Many specially designed software packages are available for carrying out discrete event simulations. The common uses of simulation models are to investigate the efficiency of proposed new systems or alterations to existing systems. A careful study of the results obtained is necessary and, used sensibly, simulation models can save a great deal of expense since experiments are far cheaper to carry out on models than in the real world.

8 USING DIFFERENTIAL EQUATIONS

8.1 INTRODUCTION

As mentioned in chapter 1, models are created with a particular purpose in mind and in many cases this purpose is to make *predictions*. These predictions may take the form of answers to such questions as 'What happens to this variable if we increase (or decrease) that other variable?' A successful model will have provided us with relationships between relevant variables from which we can obtain the necessary answers and, if it is a truly successful model, these predictions will be found to be verified when the data become available.

Very many models involve *time* as one of the variables and a very common question to ask in these cases is 'What happens to all the other variables as time progresses?' For these models we therefore need to consider *rates of change* of variables with time and it is a very likely consequence that our mathematical equations will contain derivatives. Such *differential equations* can be any of a wide variety of types, many of which cannot be solved analytically although approximate solutions can be obtained by numerical methods.

In any particular model, we arrive at a differential equation by considering the changes that occur in our variables over a finite but small time interval Δt. Dividing by Δt gives us finite rates of change and, letting Δt shrink to zero and taking limits, we obtain *differential equations.*

Such equations *can* be very difficult to solve exactly but there is no need to despair if the route to an exact solution appears closed. Much useful information about the behaviour of the solution can be gleaned from a differential equation *without* solving it. We can gain useful qualitative information by looking at such things as the magnitude and sign of the derivative for various values of the dependent and/or independent variables and, in particular, looking for values, if any, which make the derivative equal to zero. Good approximate solutions can be obtained by numerical methods and these can be just as useful as the exact analytic solutions.

Differential equations can be classified in many ways, the most fundamental being the order of the highest derivative occurring in the equation. In section 8.2, we use a first-order differential equation to model population growth, while in section 8.3, problems in mechanics are solved using a second-order equation. When our model involves several differential equations, they may be 'uncoupled', in which case each differential equation involves only one of the dependent variables, e.g.

$$\frac{dx}{dt} = A + Bx, \qquad \frac{dy}{dt} = C + Dy^2.$$

Alternatively, they may be coupled, e.g.

$$\frac{dx}{dt} = Axy, \qquad \frac{dy}{dt} = Bx + Cy.$$

Note that the independent variable t may also appear anywhere on the right-hand side of any of these equations.

In section 8.4, we have a model involving uncoupled equations, and in section 8.5, our examples involve coupled equations.

8.2 FIRST ORDER, ONE VARIABLE

Example 8.2.1: A model for population growth

If we want to create a model for predicting population growth, the variable in which we are interested is the size of the population at any time. This is the total number of living individuals in the population, ignoring the fact that they are of different ages, some are male, some are female, and so on. This total number is obviously a discrete variable but it makes the modelling easier if we pretend that it is a continuous variable $P(t)$. This will not be a serious error when $P(t)$ is large because a graph with very small discrete steps will be indistinguishable from a smooth curve.

The purpose of our model will be to predict the future population, i.e. to give $P(t)$ as a function of t. The most obvious reasons for a population to change are that individuals are born and some die but there could also be movement into and out of the area in which our population lives. So a simple model for the change in $P(t)$ during a short period of time Δt is

$$\left\{\begin{matrix}\text{increase in population}\\ \text{during time } \Delta t\end{matrix}\right\} = \left\{\begin{matrix}\text{births during}\\ \text{time } \Delta t\end{matrix}\right\} - \left\{\begin{matrix}\text{deaths during}\\ \text{time } \Delta t\end{matrix}\right\}$$
$$+ \left\{\begin{matrix}\text{immigration during}\\ \text{time } \Delta t\end{matrix}\right\} - \left\{\begin{matrix}\text{emigration during}\\ \text{time } \Delta t\end{matrix}\right\}.$$

For simplicity, we shall forget immigration and emigration. Clearly the number of births and deaths will depend on the following.

(i) The size of the interval Δt.
(ii) The size of the population at the beginning of the time interval.

The simplest assumptions to make are that a strict proportionality applies in both cases; we then have

$$\text{number of births in the time interval} = BP(t)\,\Delta t$$

and

$$\text{number of deaths} = DP(t)\,\Delta t,$$

where B and D are constants.

Our model is now

$$P(t+\Delta t) - P(t) = (B - D)P(t)\Delta t. \tag{8.1}$$

From this point we can develop our model in one of two different directions.

1 We could decide to use a *finite* time unit, say $\Delta t = 1$ year. Our variable $P(t)$ would then be discrete, being defined only at intervals of one year. Equation (8.1) is a *difference equation* relating successive P values in the sequence $P_1, P_2, P_3, \ldots,$ where $P_n = P(n)$. We could write the equation as $P_{n+1} = (B - D + 1)P_n$ and the parameters B and D are the annual percentage birth rates and death rates respectively. To calculate the values $P_1, P_2, \ldots,$ we need the value of P_0, the population size at time $t = 0$. We shall not pursue difference equation models here as our subject in this chapter is *differential* equations.
2 The alternative is to regard the population as a *continuous* variable $P(t)$ as we mentioned earlier. To proceed with our model, we first rewrite equation (8.1) as

$$\frac{1}{P(t)}\frac{P(t+\Delta t) - P(t)}{\Delta t} = B - D.$$

Letting $\Delta t \to 0$, this becomes

$$\frac{1}{P}\frac{\mathrm{d}P}{\mathrm{d}t} = B - D.$$

The expression on the left-hand side can be called the 'proportionate growth rate'. Note that it does not have quite the same interpretation as in the discrete case because we are not tied to a particular time step here. If we use a small enough time step, however, both discrete and continuous models should agree and, if $B - D$ is truly constant, equal to 0.04 say, this would approximately correspond to a net growth rate (births minus deaths) of 4% per time unit.

The solution of the differential equation is

$$P(t) = P(0)\exp[(B - D)t],$$

where $P(0)$ is the size of the population at time $t = 0$. This is a very simple model and predicts exponential growth without limit if $B > D$ or exponential decay to zero if $B < D$.

In practice, of course, we do not see populations growing exponentially without limit. The growth may follow the exponential model for some time but eventually limitations of available space or food will tend to force the population level to 'flatten out'. We can interpret this as the effect of overcrowding and/or food shortage, causing the birth rate to fall and the death rate to rise.

We can attempt to take these factors into account by allowing the proportionate growth rate $(1/P)\,\mathrm{d}P/\mathrm{d}t$ to be a function of P rather than a constant. What function of P should we take? Obviously we need one that decreases with increasing P. The simplest model satisfying this requirement is a linear function, i.e.

$$\frac{1}{P}\frac{\mathrm{d}P}{\mathrm{d}t} = \alpha - \beta P,$$

where α and β are positive constants.

Before trying to solve this differential equation, we can work out some of its implications. If the population level reaches the value α/β, then the right-hand side will be zero, which means that $\mathrm{d}P/\mathrm{d}t$ will be zero so that P will stop changing with time. Also, when we start with a population size $P(0)$ at time $t = 0$, then the right-hand side of the differential equation is $\alpha - \beta P(0)$. If this is positive, then $\mathrm{d}P/\mathrm{d}t > 0$ and the population size will be increasing at $t = 0$. Conversely, if $\alpha - \beta P(0) < 0$, we shall have a decreasing population at $t = 0$. These considerations enable us to sketch in pieces of solution curves as shown in Fig. 8.1.

We can see that, if we start with $P(0) > \alpha/\beta$, we have a decreasing population initially and this decrease will continue until $P \approx \alpha/\beta$ but *no further* because $\mathrm{d}P/\mathrm{d}t$ will be zero. Similarly, if we start somewhere below the horizontal line $P = \alpha/\beta$, our population will be increasing ($\alpha - \beta P > 0$; so $\mathrm{d}P/\mathrm{d}t > 0$) until $\alpha - \beta P \approx 0$ near the line $P = \alpha/\beta$ and the curve will never actually cross this line.

In fact the exact solution of the differential equation

$$\frac{1}{P}\frac{\mathrm{d}P}{\mathrm{d}t} = \alpha - \beta P$$

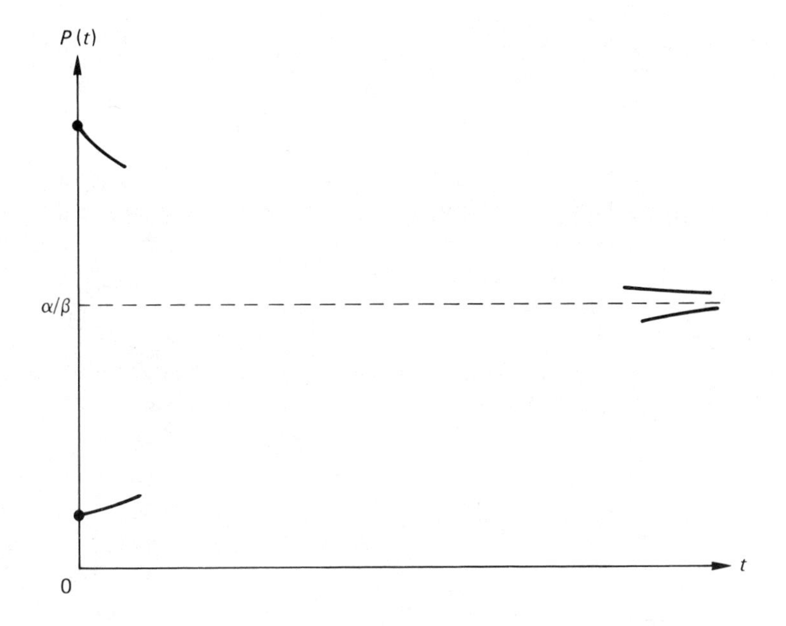

Fig. 8.1

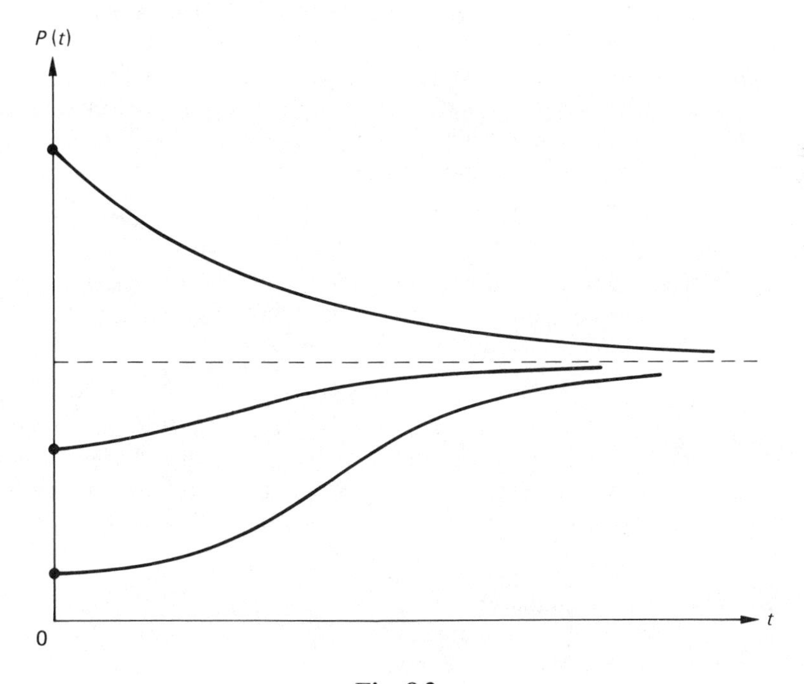

Fig. 8.2

takes the form

$$P(t) = M\left[1 + \left(\frac{M}{P_0} - 1\right)\exp(-\alpha t)\right]^{-1},$$

where $M = \alpha/\beta$ and $P_0 = P(0)$, and from this we see that the value $M = \alpha/\beta$ is never actually achieved for a finite t value. Our curves get closer and closer to the horizontal line $P = M$ as t increases. This line represents a stable population size and various solutions arising from different initial values of P converge to it, as shown in Fig. 8.2. If we had the initial value $P = M$, then of course we would have $dP/dt = 0$ for all t and the population would remain at this constant value. The population level M is sometimes referred to as the 'carrying capacity of the environment' and the curve is often called the 'logistic curve'.

EXERCISES 8.2

1 A large herd of wild animals lives in a region where the rainfall varies in a periodic manner with a maximum occurring every 4 years. The animals live by grazing and the amount of food depends on the rainfall. The net birth rate can be assumed to be proportional to the amount of food available, the maximum rate being 2% per annum. There is also a net emigration of 100 animals per annum from the region. The present size of the population is 5000 animals and predictions for the next 20 years are required. Develop a model for this problem.

2 The rate of cooling of a hot body in air is very nearly proportional to the difference between the temperature of the body and the temperature of the air. If the temperature of the air is 18 °C and a body initially at 60 °C is found to have cooled to 50 °C after 3 min, how long will it be before it cools to 30 °C? What will its temperature be after 10 min?

3 A conical tank of height 2 m is full of water and the radius of the surface is 1 m. After 8 h the depth of the water is only 1.5 m. If we assume that the water evaporates at a rate proportional to the surface area exposed to the air, obtain a mathematical model for predicting the volume of water in the tank at any time t.

4 Assume that in the absence of fishing the population level of a species of fish would follow the same model as discussed in section 8.2. The 'maximum sustainable yield' Y is the maximum amount of fish biomass that can be caught annually without causing the fish stocks to decrease. Obtain a model for predicting Y from the other parameters of the problem.

8.3 SECOND ORDER, ONE VARIABLE

No text which sets out to be a first course in mathematical modelling would be complete without some treatment of physical models involving the use of Newton's laws of motion. In a sense, in physical modelling, the 'formulation' part, normally a key feature in the process as described in chapter 3, is reduced as we are relying on Newton's laws to provide us with a ready-made framework on which to base the mathematical model. Close to the Earth's surface, Newton's laws are accepted as giving a reliable description of the behaviour of mechanical systems. We have certainly no intention of challenging Newton here; so it appears as if the student modeller is in for an easy time! If this were completely true, it would be a pity since it is usually the formulation stage which gives the greatest challenge, i.e. you have the opportunity of building up relations and equations between variables based on your own assumptions or collected data.

In considering a model for a physical or mechanical system, we shall be seeking an *equation of motion* based on Newton's second law. We do not attempt to interfere with this law; however, it often turns out that not all the constituent parts of the equation of motion will be beyond discussion. For example, any situation involving consideration of air resistance, frictional effects, size of moving object, etc., will need our modelling judgement as to their inclusion in the model.

While we are discussing preliminaries, it is important to point out that mathematical modelling and models are common in the 'real' world where there is a physical or engineering system under investigation. It is not a question of tinkering with problems found in textbooks on classical mechanics. Modelling a physical system is widely carried out in industries involving the transportation or flow of a substance: gas, coal, steel, electricity, nuclear power, etc. Also the modelling of wave behaviour is important in radar and submarine detection work.

In section 8.2, we looked at a model for population growth and this will be followed in section 8.5 with a harder population model and also a battle model. These could be classified as biological models, although it can be a mistake to classify too rigidly. A *physical systems model* is intended to refer to situations involving *physics* such as mechanics, heat transfer, sound waves and electricity. Thus we are concerned with dropping, swinging, throwing, hitting, heating and cooling, flowing, diffusing and possibly even permeating.

As we have already said, the development of models for physical systems relies very heavily on the application of Newton's second law of motion. This can be stated as follows:

$$\begin{Bmatrix} \text{net external} \\ \text{force} \end{Bmatrix} = \begin{Bmatrix} \text{rate of change in linear} \\ \text{momentum of a system} \end{Bmatrix}$$
$$= \text{mass} \times \text{acceleration}$$

(assuming constant mass).

Now this section deals with second-order differential equations, and it is the concept of acceleration which gives us the second derivative term. We shall not describe here the full details of theoretical mechanics since there are many good texts on the subject (see, for example, D. G. Medley, *An Introduction to Mechanics and Modelling*, Heinemann, 1982). Instead we concentrate on the creative theme embodied in modelling. Two examples will now be investigated which contain this theme. In both of the resulting models, we shall find that the mathematical representation results in a *second-order differential equation*, and this contrasts with the models in section 8.2 which all resulted in *first-order differential equations*. Also both examples contain aspects of mechanics that are found in first-course modelling books, but the examples also contain interesting features which you may not have met before.

Example 8.3.1

This example is set in the context of a recent request made to one of the authors by a solicitor acting for a client in the court of appeal. The client was a convicted murderer suspected of having jumped from a high window to avoid detection. The legal problem centred on the client's claim that, had he been the person jumping, then he would have severely injured himself owing to previously known personal leg joint weakness. The mathematical problem was to estimate the impact speed in these circumstances to see whether it was likely that the person could then get up and run away!

Problem statement

The particular question posed was 'If someone fell a distance of about 30 ft, what would the impact speed be?'

Formulate a mathematical model

Here is our chance to make some modelling judgement about the nature of falling human bodies: is it free fall or air-resisted fall? Does the size of the falling body matter in this case? Assuming that air resistance does have an important effect, how are we going to assess it in our model? Following our usual procedure we first list the factors involved (Table 8.1).

Now there are useful general connections between x, v, t and f using differential calculus which can be applied in this context:

$$v = \frac{dx}{dt},$$

$$f = \frac{dv}{dt} = \frac{d}{dt}\left(\frac{dx}{dt}\right) = \frac{d^2x}{dt^2}.$$

Table 8.1

Description	Symbol	Units
Distance fallen	x	m
Velocity	v	m s^{-1}
Time	t	s
Acceleration	f	m s^{-2}
Mass	m	kg
Gravitational constant	g	9.8065 m s^{-2}
Gravitational force	mg	N
Air resistance force	R	N

Note that, for convenience, $\mathrm{d}x/\mathrm{d}t$ and $\mathrm{d}^2x/\mathrm{d}t^2$ are written as $\dot{x}$ and $\ddot{x}$, respectively. Also

$$f = \frac{\mathrm{d}v}{\mathrm{d}t} = \frac{\mathrm{d}v}{\mathrm{d}x}\frac{\mathrm{d}x}{\mathrm{d}t} = v\frac{\mathrm{d}v}{\mathrm{d}x}.$$

A diagram usually helps in situations such as this; so, before applying Newton's second law, we shall give one, labelling it using the above notation (Fig. 8.3). Note that O is the point from which the body starts its motion

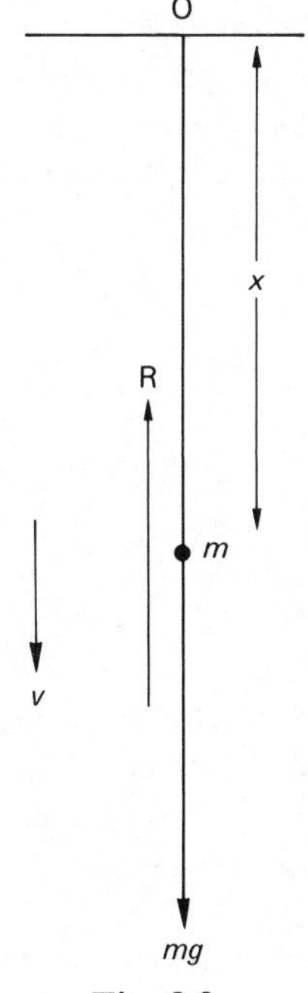

Fig. 8.3

and the variable x denotes the distance fallen vertically downwards in time t. All mechanics problems need a fixed frame of reference and, as the motion is vertically downwards, we have a one-dimensional problem.

The forces acting have magnitudes mg and R in the directions shown on the diagram; so the net external force on the body has magnitude $mg - R$. Applying Newton's law (in a downward positive sense), we get

$$mg - R = m\frac{\mathrm{d}^2x}{\mathrm{d}t^2} = mv\frac{\mathrm{d}v}{\mathrm{d}x}.$$

This is our model. It is not completely formulated as R has to be specified. Note also the mg term at the beginning of the equation, which represents the force due to gravity. The acceleration due to gravity can be measured experimentally in vacuum to be constant near the Earth's surface. This constant value is approximately 9.8065 m s^{-2} or 32.2 ft s^{-2}. Thus,

$$\begin{aligned}\text{force due to gravity} &= \text{mass} \times \text{acceleration}\\ &= mg.\end{aligned}$$

To proceed further, we need an expression for R. Air drag is a common enough experience whether running, cycling or even walking. (Is it the same thing as considering the effect of the wind?) A moment's thought tells us that R will not be *distance* or *time* dependent (why not?) but will be velocity dependent, i.e. the faster you go the greater the air drag. So we model air resistance as $R = kv$, provided that R is directly proportional to v.

Suppose that a more complicated relation is proposed, such as $R = kv^2$ or, more generally, $R = kv^n$. How are we to know which is the best model? The answer to this is that it depends what sort of object is being dragged. For the moment, let us take the general relation $R = kv^n$, and the equation of motion then becomes

$$mg - kv^n = mv\frac{\mathrm{d}v}{\mathrm{d}x}.$$

We can divide through by m, redefine the constant of proportionality k as $K = k/m$ and so produce a neat mathematical model:

$$g - Kv^n = v\frac{\mathrm{d}v}{\mathrm{d}x}.$$

However, what is to be done about the index n? A number of attempts have been made to decide this, and also the value of K which now contains dependence on the mass m of the object. These decisions are important in modelling; so we must not fudge the issue. We shall use the following information from books on mechanics.

(a) For a small compact object such as a stone, air resistance is directly proportional to speed, i.e. $n = 1$.

(b) For large bulky objects such as a human body, air resistance is directly proportional to the square of the speed, i.e. $n = 2$, and this is what we shall use in our model. This makes the mathematical model easier to solve than if n were taken to be, say, 1.347.

The value of K has also to be worked out so that the magnitude of the air resistance can be correctly estimated. This can be done by looking up a textbook on mechanics or, better still, using the concept of 'terminal speed' (see section 4.4). The terminal speed is the ultimate constant speed acquired by a body moving freely in a medium such as air. It is reported that the terminal speed of a human is about 120 miles h^{-1}. In this situation, there is no acceleration (gravitational pull balances air resistance) and from the differential equation this means that $g - Kv^2 = 0$. Converting 120 miles h^{-1} to the equivalent value in metres per second and substituting $g = 9.8065$ gives $K = 0.003\,41$. Thus, our model for the motion of a body falling out of a window is represented by the first-order differential equation

$$g - 0.003\,41v^2 = v\frac{dv}{dx}.$$

Obtain the mathematical solution

Provided that $n = 2$ (as here) or $n = 1$ the equation can be integrated directly. It does not worry the modeller, however, to be told that $n = 1.347$ since numerical methods can be used or perhaps a simulation package which deals with general differential equation models. The mathematical solution for $n = 2$ is

$$x = \frac{1}{2K} \ln\left(\frac{g}{g - Kv^2}\right). \tag{8.2}$$

If we want the relation between distance fallen and time elapsed, then further integrations give

$$x = \frac{1}{K} \ln[\cosh(t\sqrt{gK})] \tag{8.3}$$

($K = 0.003\,41$). The corresponding results for the dropping of a stone (where $n = 1$, and a different K value is then needed) are

$$x = \frac{g}{K^2} \ln\left(\frac{g}{g - Kv}\right) - \frac{v}{K} \tag{8.4}$$

and

$$x = \frac{g}{K}\left(t + \frac{1}{K}\exp(-Kt)\right) - \frac{g}{K^2}. \tag{8.5}$$

Table 8.2

x/ft	0	5	10	15	20	25	30	35
v/miles h^{-1}	0	12.20	17.22	21.03	24.23	27.02	29.52	31.8

Equation (8.2) can be used to tabulate x and v. The original question about the man jumping out of a window can now be answered. We must convert the units to *feet* and *miles per hour* since that is how the solicitor will want them. Using the conversion factors from chapter 4, Table 8.2 can be drawn up.

The particular distance jumped or fallen was said to be 30 ft; so the model predicts that the speed on hitting the ground is 29.52 miles h^{-1}.

Interpret the mathematical solution

First note that to declare a value for the impact speed correct to two decimal places is not sensible for the following reasons.

(a) The precise distance dropped is not known, only 'about 30 ft'.
(b) The result is best given correct to the nearest whole number for simplicity.

The advice produced from our model is that, if you fall about 30 ft, you hit the ground at about 30 miles h^{-1}. This could be compared with being in a car crash at 30 miles h^{-1}; so we expect some injury. Also, in parachute training, instructors often say that the impact will be like falling off a 12 ft wall, which from Table 8.2 is equivalent to an impact speed of 19 miles h^{-1}.

It therefore seems reasonable to suggest that falling 30 ft would give some injury, the impact speed being about 50% higher than that for a parachutist. However, this does not take into account the nature of the ground where the landing takes place. Perhaps a soft landing in a flower bed might allow the criminal to get away unhurt. Can we include these considerations in our model?

Injury on impact will be related to the magnitude of the average force experienced on hitting the ground. We can make a numerical estimate of this using the principle that

$$\text{force} \times \text{distance} = \text{work done.}$$

In this case the force concerned will be the average force experienced over the time interval during which the person is sinking into the soil and the distance will be the depth to which he sinks. The work done is equated to the total kinetic energy on impact, given by 0.5 × mass × (speed)^2. Knowing the mass and speed, if we also know the depth of sinking into the soil, we can calculate the average force experienced. Our advice to the solicitor is to examine the ground below the window.

Further thoughts

1 If air resistance is neglected, what difference does this make to the model and what answer is then obtained?
2 If examination of the soil beneath the window reveals footprints 10 cm deep, what can we deduce?

Example 8.3.2

You will probably be familiar with the model for a simple pendulum. Here we shall deal with a more interesting real-life situation. Suppose that a heavy mass is being swung against an old high wall on a demolition site. The situation is quite common where old buildings are being knocked down prior to site redevelopment. The heavy mass is supported from the end of a crane cable and is made to swing by the crane driver making the crane jib move backwards and forwards. The objective is to make the mass strike the wall as hard as possible so that parts of the wall fall down. The crane driver can also pay in or pay out the cable as the mass is swinging backwards and forwards.

Problem statement

A single wall of height 12 m is to be demolished. The crane jib is 15 m above ground level and the driver decides to place the cab about 10 m from the wall. Initially the heavy mass hangs at rest with the cable extended to a length of 8 m. The driver now operates the crane with the intention of producing a suitable swing of the heavy mass (30 kg) so that it will strike the wall with the greatest impact velocity possible. The problem is to find this swing. The driver also wishes to strike the wall as high up as he can since clearly the wall will be knocked down most easily by hitting it near the top. (Why?)

Formulate a mathematical model

Unlike Example 8.3.1 the path described by the mass P can only be expressed using *two* dimensions. It is nevertheless possible for the eventual mathematical model to be expressed by a second-order differential equation in one variable. To fix ideas, a diagram is needed (Fig. 8.4). A fixed frame of reference is required and the problem variables must be related to this frame. A suitable notation for the variables is marked on the diagram, and a list is given in Table 8.3.

Apart from neglecting air resistance and wind effects, one key assumption

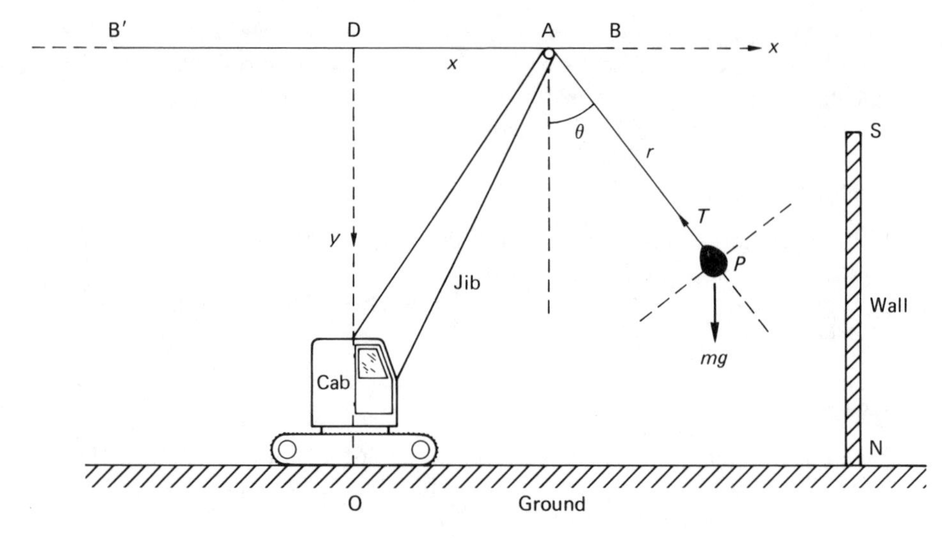

Fig. 8.4

Table 8.3

Description	Symbol	Units
Cable length	r	m
Jib top displacement	x	m
Maximum jib top displacement	a	m
Angle between cable and vertical	θ	deg
Swing frequency	ω	s^{-1}
Mass	m	kg
Tension in crane cable	T	N
Gravity force	mg	N
Cable pay-out speed	u	m s^{-1}
Time	t	s
Impact speed	v	m s^{-1}

will be made concerning the motion of the jib top. We shall assume that it moves in a straight line perpendicular to the wall when activated by the driver. As its movement will be relatively small compared with the height of the wall, this seems a reasonable approximation. This means that the cable AP will move in a vertical plane perpendicular to the wall. Thus, from the diagram, BD, OD and ON are fixed lines.

The mathematical model will be based on applying Newton's second law to describe the motion of the mass P. This requires the acceleration of P to be expressed in terms of the problem variables x, r and θ. With respect to the point A, P has *polar coordinates* (r, θ). The radial and transverse components of velocity and acceleration in terms of polar coordinates are well known (see, for example, D. G. Medley, *An Introduction to Mechanics and Modelling*, Heinemann, 1982). Let us mark them on a separate diagram (Fig. 8.5) for easy reference.

As the point A is itself moving horizontally with displacement denoted by x, then an additional horizontal acceleration component is added. This is shown in Fig. 8.4. There will be two forces acting on P, again labelled in Fig. 8.4. These are the gravity force mg and the tension in the cable denoted by T. (We assume calm local conditions; so wind and air resistance are neglected.)

In order to knock down the wall, we imagine the crane driver causing the jib arm to move back and forth so that oscillations are transferred to the point P. In two dimensions, when Newton's second law is applied, we can expect two equations of motion, but a moment's thought tells us that we are not really interested in bringing into the mathematical model the value of the *cable tension.* Consequently consider the equation of motion in a direction perpendicular to the cable. The resulting acceleration component in this direction will be made up from the polar terms $2\dot{r}\dot{\theta} + r\ddot{\theta}$ relative to A, together with the term $\ddot{x}\cos\theta$ from the motion of A resolved in the direction perpendicular to the cable. The equation of motion is

$$-mg\sin\theta = m(2\dot{r}\dot{\theta} + r\ddot{\theta} + \ddot{x}\cos\theta). \tag{8.6}$$

As expected, this is a second-order differential equation in the variable θ but also containing the variables x and r. However, these can be considered as inputs to the model as they are under the control of the driver.

Now the crane driver has to activate the system by swinging the jib backwards and forwards. This can be *modelled* by taking the variable x to

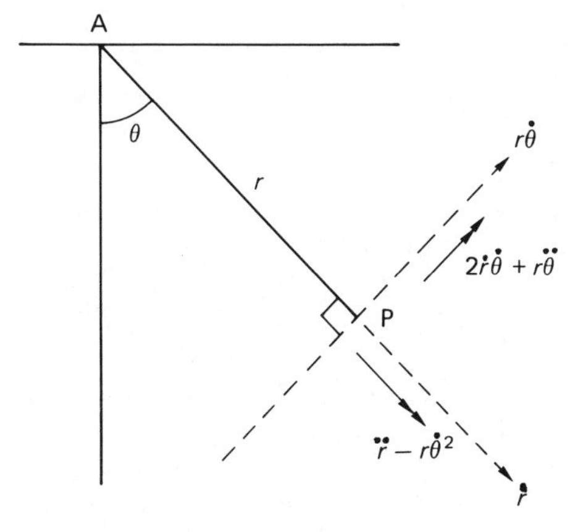

Fig. 8.5

have a periodic form of the type

$$x = a\begin{cases}\sin(\omega t),\\ \cos(\omega t),\end{cases} \tag{8.7}$$

according to the starting conditions that we select. This function is an obvious choice for representing the physical effect that we are seeking. By suitably choosing values for the *amplitude* a and the *frequency* ω, we can model the desired effect.

We now consider the pay-out of the cable. We might suppose that the initial length of the cable is a certain value, say r_0, and, if we pay out the cable at a steady rate u, then a model for the variable r is

$$r = r_0 + ut. \tag{8.8}$$

By substituting from equations (8.7) and (8.8) for x, r and their derivatives the differential equation for θ is

$$\ddot{\theta} = \frac{-g\sin\theta - 2u\dot{\theta} + a\omega^2\sin(\omega t)\cos\theta}{r_0 + ut}. \tag{8.9}$$

Note that other forms of equation (8.9) will be obtained if alternative models are used for x and r. For example using equation (8.8) means that the pay-out speed of the cable is always u which is counter to the original condition that the mass at P is initially at rest.

Equation (8.9) is considerably more complicated than the one we had in Example 8.3.1. We could decide to simplify the model, but in any case the mathematical solution will probably require a numerical method or perhaps the use of a *simulation package.*

Another interesting factor here is that, by solving the above equation, we shall not exactly provide information that a crane driver can use or indeed is actually interested in. He is concerned with where to position his crane and also what speeds can be generated by the swings that he is setting up.

This is an example where the mathematical results obtained from a difficult differential equation, although interesting in themselves, are not the primary concern of the mathematical model. The questions of the speed of the mass and the position of the crane can be dealt with by returning to the diagram and considering the relevant variables. A new diagram (Fig. 8.6) showing the velocity components is helpful. (Note that $\dot{r}$ and $r\dot{\theta}$ are components of P's velocity relative to the point A.)

The position of the mass with respect to the coordinate origin O can be easily worked out from MP and OM:

$$\text{MP} = x + r\sin\theta,$$

$$\text{OM} = h - r\cos\theta.$$

We are interested in the speed with which the mass hits the wall. More particularly, since damage is to be inflicted by the blow, it is the horizontal

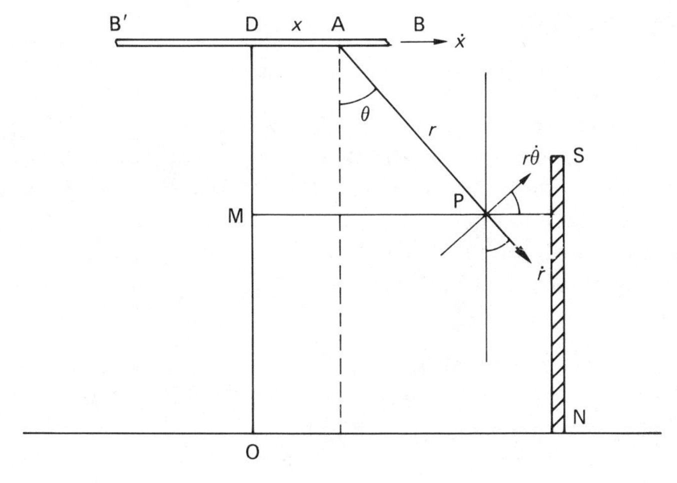

Fig. 8.6

velocity component that has to be calculated, given by

$$v_{\text{horiz}} = \dot{x} + \dot{r}\sin\theta + r\dot{\theta}\cos\theta.$$

Obtain the mathematical solution

We are now ready to investigate the solution of the mathematical model. We must decide where to start the jib oscillations from, and what energy to put into the system. We also have to consider whether to pay out the cable and at what rate. Most importantly of all, we have to know the dimensions of the wall to be demolished and how close to this wall we can position the base of the crane. This means that the parameters a, u and ω have to be chosen.

To take a particular case, we shall use the following data, but you may like to try other values: $a = 3.0$ m, $u = 0.3$ m s^{-1}, $\omega = 1.0$ s^{-1} and $r_0 = 8.0$ m. Initially P hangs below the point D (see Fig. 8.6).

Interpret the mathematical solution

With the above data, we find that the wall will be struck after one swing with a speed whose horizontal component is 9.0 m s^{-1} at a height of about 11 m. There is a final comment to make about the use of this model, and that is that *the wall will be knocked down according to the size of the blow inflicted.* The value of the mass m has not been needed so far, but the magnitude of the blow will be given by the horizontal momentum at impact, i.e. $m \times v_{\text{horiz}}$. We also do not know the strength of the wall, but we assume it to be only so strong that this method of demolition is suitable.

EXERCISES 8.3

1 *Newton's laws: fact or fiction exercises* Often when we come to apply Newton's laws in mathematical models, the first problem is to be sure that we have understood the mechanical principles involved. If your modelling has been supported by, say, a parallel course in dynamics and statics or some suitable alternative, then you will probably have no trouble in marking in the correct forces and accelerations in a given physical situation. If you are more uncertain, try the following exercises.

Classify the following as true or false.

(a) A force always produces motion.
(b) A body always moves in the same direction as the force acting on it.
(c) If no force is acting on a body, then the body must be at rest.
(d) The velocity produced in a moving body is directly proportional to the force acting.
(e) When a ball is thrown vertically upwards (Fig. 8.7), then at A (the highest point reached), the velocity and acceleration are

(i) both zero,
(ii) one of them is zero (which?), or
(iii) neither is zero.

(f) When a ball is thrown as a projectile (Fig. 8.8), then at B (the highest point)

(i) the horizontal velocity component is zero,
(ii) the horizontal acceleration component is zero,
(iii) the vertical velocity is zero, or
(iv) the vertical acceleration is zero.

Fig. 8.7

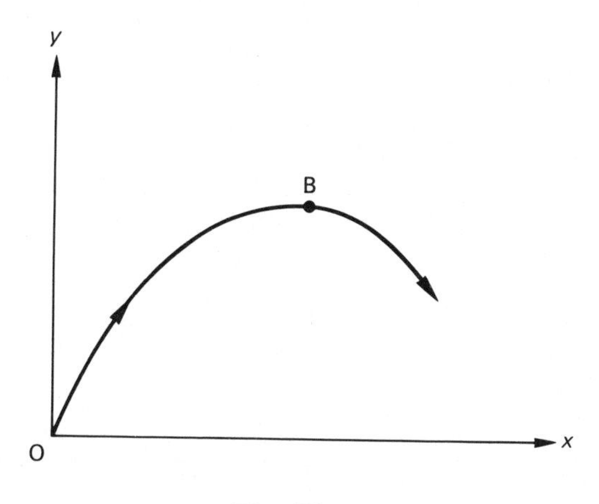

Fig. 8.8

2 Now try to answer the following problems.

(a) When you push a door shut, why is it easier to push near to the door handle than near to the hinge?
(b) Why is it easier to pull a heavy object (i.e. a lawn mower or a sledge) along than to push it?
(c) How does the fairground wall of death work?
(d) Why do astronauts experience weightlessness in a capsule a few hundred miles up from the Earth?

3 On some 'private' estates and also within the grounds of a hospital, the authorities install a series of small ramps across the road to prevent motorists from exceeding a certain speed limit. (The ramps are sometimes called 'sleeping policemen'!) The effect of driving too fast over such a ramp will be to cause the driver some discomfort and he or she may hit the car roof. Investigate the situation when a car moves over the ramp. Consider the effect of the springs of the vehicle which is travelling over a flat road. The springs will be compressed into some steady-state condition. A car suspension system also uses dampers to prevent continuing oscillation.

Using the very simple model of a car shown in Fig. 8.9, formulate a differential equation for the displacement of the driver. This will be a second-order equation and is not too difficult to solve. (You will need to use Hooke's law to model the car spring: the force of tension or compression is directly proportional to the amount that the spring is stretched or compressed. Also assume the damper force is proportional to the rate of change of displacement.) The data are as follows: the mass of the car is 1000 kg, the spring stiffness is 5×10^4 N m^{-1}, the damper coefficient is 9×10^4 N m^{-1}, and the ramp is about 1 m across and its height is about 30 cm.

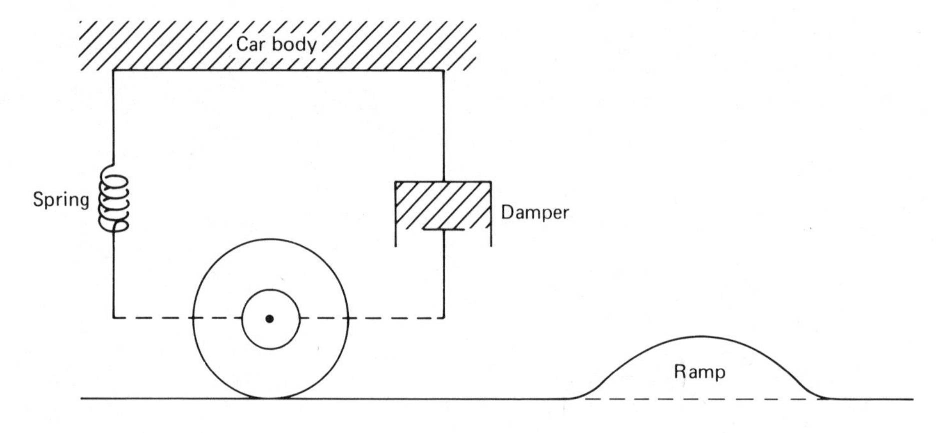

Fig. 8.9

4 A common feature of all public buildings is some mechanical device which will automatically close a door after someone has passed through. Can you set up a model to explain how the effect is usually for the door to close fast when swung wide open, but to close more gently as it moves to the shut position? This seems quite a large task, and so we shall suggest two systems for investigation: one simple (case A) and the other more complicated (case B).

Case A is depicted in Fig. 8.10 which shows a simple 'arm' which presses onto the door with a certain force due to a coiled spring in the spring unit. The idea is that, with the door almost shut, the pressure from the arm has reduced almost to zero. As far as the motion of the door is concerned, you will need to regard its motion as that of a rigid body. This means that we shall need to know the 'moment of inertia' of the door about the vertical hinge axis (see, for example, D. G. Medley, *An Introduction to Mechanics and Modelling*, Heinemann, 1982).

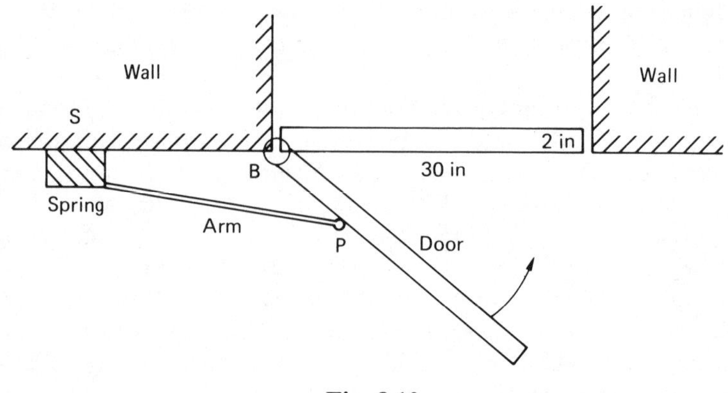

Fig. 8.10

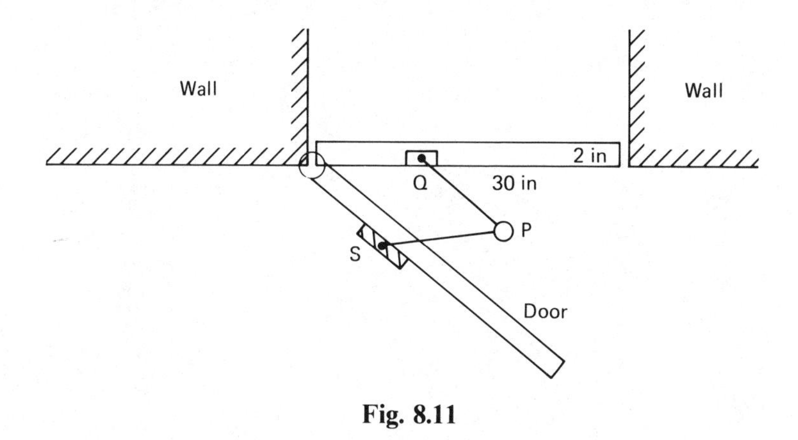

Fig. 8.11

Case B is the more common two-armed version shown in Fig. 8.11. Here contact between the mechanism and the door and door frame are at fixed places and there are two freely joined arms which cause the door to close.

5 A length of chain is hanging over the edge of a horizontal deck. The weight of the part hanging vertically down causes the whole length to slide off. Motion is slowed by the friction between the chain and the deck. Formulate a mathematical model to predict the motion of the chain.

8.4 SECOND ORDER, TWO VARIABLES (UNCOUPLED)

When a body moves in two dimensions, we need two coordinates to specify its position at a given time. Effectively we apply Newton's second law in two directions to obtain two equations of motion. Usually these equations are uncoupled in the sense that each coordinate variable appears in only one equation. The resulting mathematical formulation will be two simultaneous second-order differential equations. Examples of this situation occur when we are modelling projectile motion. As an example which contains features a little different from the standard projectile, consider the following 'fireman' problem.

Example 8.4.1

Context

A fireman is directing a jet of water into a window of a burning building. Owing to the heat, and the danger of a wall's collapsing, the fireman wishes

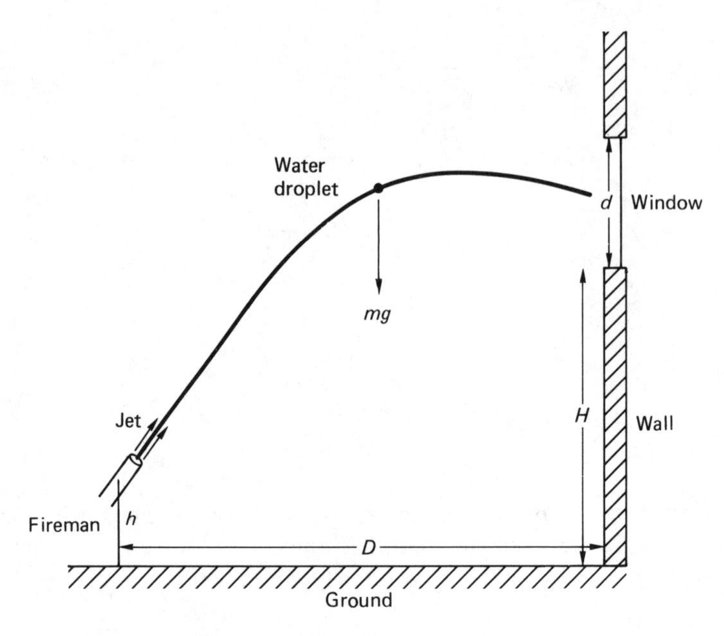

Fig. 8.12

to stand as far back as possible, while still being able to direct the water jet through the window. The scene is illustrated in Fig. 8.12.

Problem statement

In mathematical terms, the problem objective is to find the maximum value of the distance D so that the jet still reaches the window.

Formulate a mathematical model

Some of the quantities in Table 8.4 are more important than others. Depending on the sophistication of the model, we may decide to ignore certain features. To make a start, some assumptions are listed as follows.

(a) The water jet is effectively a stream of particles.
(b) We can neglect air resistance, cross-winds, change in water pressure, etc.
(c) The size of the window is neglected in the first model.

It is convenient to draw another diagram (Fig. 8.13) in which the forces and velocities can be shown.

As we have hinted at the start of this section, we need to obtain *two* equations of motion. One of these will represent the horizontal motion and the other the vertical motion. Newton's second law can be applied in each direction separately and, in doing so, we get a pair of *uncoupled*

Table 8.4

Description	Symbol	Units
Initial velocity of water jet	u	m s^{-1}
Height of window	H	m
Size of window		
Width	w	m
Height	d	m
Forces acting		
Gravity ($g = 9.8065$ m s^{-2})	mg	N
Air resistance	R	N
Height of fireman	h	m
Initial angle of water jet	α	deg
Area of cross-section of water jet	A	m^2
Distance of fireman from base of wall	D	m
Time	t	s
Coordinate variables	x, y	m
Speed	v	m s^{-1}

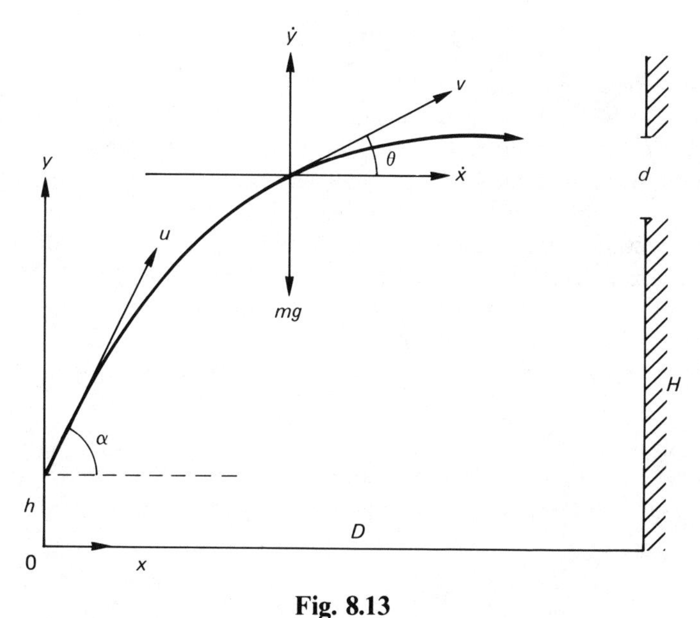

Fig. 8.13

equations, i.e. the dependent variables do not appear in both equations. This uncoupledness will help us to obtain the solution to the model. In the next section, we see a situation where the model involves *coupled* differential equations with a resulting different approach to the solution. Remembering that Newton's second law states that force = mass × acceleration, we now apply this law to get our two equations: for horizontal motion

$$0 = m\frac{d^2x}{dt^2} \tag{8.10}$$

and, for vertical motion,

$$-mg = m\frac{d^2y}{dt^2}. \tag{8.11}$$

Note that we have inserted m into the equations without saying what m represents. We can imagine a small group of droplets sufficiently close to be taken together as a particle of mass m. The jet of water can be regarded as a stream of such particles. The point about uncoupledness can be seen in equations (8.10) and (8.11) since the dependent variables are x and y.

Obtain the mathematical solution

Two straightforward integrations of these simple ordinary differential equations give

$$x = ut\cos\alpha, \tag{8.12}$$

$$y = ut\sin\alpha - \tfrac{1}{2}gt^2 + h \tag{8.13}$$

(do not forget the height of the fireman which is h).

This is nice and familiar, but what are we supposed to be trying to solve? This was how far away the fireman can stand and still direct his jet through the window. A moment's thought tells us that we are not really interested in *how long* it takes for the water to reach its destination but are concerned with the distances and angles involved. (Anyone in a burning building wishing to be rescued certainly will be interested in how long the water takes to arrive, but that is rather different.) So, if time t is eliminated between equations (8.12) and (8.13), then

$$y = x\tan\alpha - \frac{gx^2}{2u^2\cos^2\alpha} + h.$$

This equation applies to all points on the jet. In particular, taking the jet to pass through the point $x = D$, $y = H$ (corresponding to the bottom of the window), then

$$H = D\tan\alpha - \frac{gD^2}{2u^2\cos^2\alpha} + h. \tag{8.14}$$

Now what interpretation do we put on this equation? On reflection we realise that H, h, g and u are fixed, so that D is the variable to be maximised over changing α. Regarding α as a continuous variable, we can differentiate in the usual way and put $dD/d\alpha = 0$.

This results in the condition $u^2/g = D_M \tan\alpha_M$, where D_M and α_M are the optimum values sought. After some manipulation, we arrive at the equations that we want as the *mathematical* solution of the model:

$$\sin^2 \alpha_M = \frac{u^2}{2[u^2 - g(H-h)]}, \tag{8.15}$$

$$D_M = \frac{u\sqrt{u^2 - 2g(H-h)}}{g}. \tag{8.16}$$

Note that equation (8.16) is valid only for $u^2 > 2g(H-h)$. Why do you think this condition is necessary?

Interpret the solution

Can this result be validated? The following data are supplied and are substituted into equations (8.15) and (8.16). Given that $u = 20$ m s^{-1}, $h = 1.5$ m and $H = 15$ m, then it is easy to calculate D and α to get $D_M = 23.7$ m and $\alpha = 59.8°$.

Using the model

This model is capable of further development since there are several factors that we have neglected or not investigated which might affect the usefulness of the results.

(a) At what angle does the water jet enter the window?
(b) The size of the window has not been brought into the model.
(c) How sensitive are the results to changes in u, h and H?
(d) A water jet will spread out as it travels from the pipe; can we say anything about this spread?
(e) Air resistance acting on the water flow has been ignored—was this justified?

The modifications necessary to deal with all these situations are left for you to incorporate, but we can see that (b), for example, will affect the validity of equation (8.14) since it is not necessary for the water jet to pass through the 'coordinate' point (D, H). The spread of the water jet is actually easy to quantify by considering 'mass conservation' in Fig. 8.14.

As the cross-section of pipe is A (m) and the initial velocity of the water jet is u (m s^{-1}), the initial volume flow equals uA (m^3 s^{-1}). If at some subsequent stage the velocity is v with cross-section A', then by the conservation of mass we have $uA = vA'$. Taking the diameter of the pipe held by the fireman to be 10 cm, then since $u = 20$ (m s^{-1}) we have $vA' = 0.157$. Thus, if v is calculated at some point, in particular as the water is about to enter the window, then we can say something about the spread (do you think that the jet is still circular in its cross-section?). On the assumption that the fireman wishes to

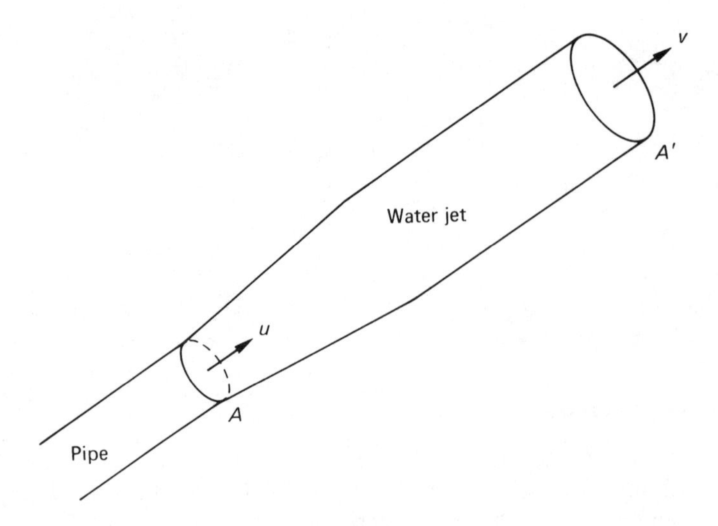

Fig. 8.14

get all the water into the open window, this analysis will reduce the variation in D consistent with hitting the target.

EXERCISES 8.4

There are a number of modelling problems involving throwing, each of which can be tackled using a similar mathematical model to that used in Example 8.4.1. Sports and games are a rich source for such investigations and we shall give three situations here.

1 *Football throw-in* A throw-in is taken from a point roughly in line with the penalty spot with the intention of landing the ball on the 'striker's' head. The team coach has said that running up before throwing is necessary in order to make a long throw. Unfortunately there is a wind blowing directly across the field towards the thrower. Also a football is affected by air resistance, which is taken as directly proportional to the speed of the ball. Your job as modeller is to construct the equations of motion which define the path of the ball and to decide whether the striker is likely to have a chance of getting his head to the ball as it comes across. The data are as follows: the run-up speed of the thrower is 4 m s^{-1}, the speed of the throw (relative to the thrower) is 15 m s^{-1}, the wind velocity varies between 0 m s^{-1} and 15 m s^{-1}, and the air resistance constant (per unit mass) is 0.003 s^{-1}.

2 *Basketball shot* When a foul occurs in a game of basketball, the attacking team may be awarded a free shot in which a player attempts, unhindered, to throw the ball into the basket from a certain position. In a way this is quite like the fireman problem, but the entry of the ball into the basket is now crucial since, as the thrower soon finds out, the ball is apt to strike the basket rim and to bounce out again! Obtain the two equations of motion again. Neglect air drag but this time the angle of flight of the ball is needed so that we can examine whether it will actually pass into the basket. Clearly the height of the thrower now matters and also the height of the gymnasium must place some restriction on the ball's trajectory. The data are as follows: the basketball diameter is 25 cm, the basket hoop diameter is 45 cm, the height of the gymnasium is 6.5 m, the distance of the throwing point from the basket is 4.6 m, and the height of the basket above the gymnasium floor is 3.05 m. Often a successful throw is achieved by rebound from the backboard, placed 0.5 m behind the centre of the basket rim. Can you incorporate this feature into your model?

3 *Putting the shot* In this problem, we wish to maximise the distance thrown by a shot putter, where the throw range is at a level *different* from the level of the throw projection point. The world record is currently around 70 ft. The question for the mathematical modeller is to offer advice regarding the angle of throw. The projection speed remains about the same for all throws, the ambient conditions in a given competition can be discounted since shot putting is a fairly local activity, and so the variables reduce to the following.

(a) The angle of throw.
(b) The height of projection.

8.5 SIMULTANEOUS COUPLED DIFFERENTIAL EQUATIONS

In many models the variables affect each other and the equation for the rate of change of one variable is very likely to involve the other variables as well. This leads to systems of coupled differential equations which as a rule are rather difficult to solve analytically. If time does not appear explicitly on the right-hand side of our equations, we can simply divide one equation by another to obtain an equation not involving t. Suppose for example that we have the following equations:

$$\frac{\mathrm{d}y}{\mathrm{d}t} = 3x - y,$$

$$\frac{\mathrm{d}x}{\mathrm{d}t} = x + y.$$

Division gives

$$\frac{dy}{dx} = \frac{3x - y}{x + y}.$$

We could now go on to solve this differential equation, which would give us a set of curves in the $x-y$ plane. This is sometimes called a 'phase plot'. In this diagram, time does not appear explicitly but is implied in the sense that a particular curve in the phase plot shows how x and y change together as time passes. We can think of the state of our system as a moving point starting from a point determined by the initial values $x(0)$, $y(0)$, and subsequently moving along a solution curve of the differential equation, usually referred to as the 'trajectory'. This can be very useful for determining the evolutionary behaviour of the model and for investigating how different starting values may lead to different final results.

What this diagram cannot do, however, is tell us how much time is required for the system to evolve from one point to another. To answer such questions, we have to go back to the original differential equations. We *may* be able to substitute for one variable from one equation into another and to solve the resulting differential equation. Alternatively, there are numerical techniques available for solving systems of coupled differential equations. If we use a numerical method, we also have to select starting values for our variables and usually the time step. We then obtain a print-out of the evolution of the model starting from the chosen initial conditions. To find out what happens from other starting conditions, we have to do separate runs of the model.

It is also worth considering what information we can get *without* solving the differential equations. In particular, it is often instructive to look at values of the variables (if any) which make the derivatives equal to zero. These are known as the 'equilibrium points' and they can have any of a number of different stability properties. We can also look for restrictions on the values of the variables implied by the signs of the derivatives.

Example 8.5.1: Modelling a battle

Context

The X army is about to attack the Y army which has only 5000 troops while the X army has 10 000. The Y army, however, has superior military equipment which makes each Y soldier 1.5 times as effective as an X soldier.

Objective

We wish to develop a mathematical model for the resulting battle and use the model for the following purposes.

1 To predict which army will win.
2 To estimate how many troops of the winning army will be left at the end.
3 To calculate how many troops the losing army would have needed initially in order to win the battle.

Formulate a mathematical model

Before getting around to developing a model for this specific problem, let us look at the assumptions we can reasonably make when modelling a battle between two opposing forces. Our variables are $x(t)$, the number of X troops alive at time t, and $y(t)$, the number of Y troops alive at time t. As in section 8.2, we realise that these should be discrete variables but we make things easier by blurring the distinction. We also assume that, at any time, all the troops of both sides are either alive and fighting or dead, i.e. we do not consider any prisoners or wounded.

It seems reasonable to assume that, in an interval of time Δt, the number Δx of X troops killed will depend on the length of the interval Δt and on the number of Y troops opposing the X army at the beginning of the time interval. Assuming simple proportionality, we have

$$\Delta x = -ay\,\Delta t,$$

where a is a constant representing the effectiveness of the Y army. To be explicit, a equals the number of X soldiers killed by each Y soldier in one time unit and can be referred to as the 'kill rate'. We have a similar relation for the Y army:

$$\Delta y = -bx\,\Delta t.$$

Letting $\Delta t \to 0$, we derive the two differential equations

$$\frac{\mathrm{d}x}{\mathrm{d}t} = -ay, \qquad a > 0 \tag{8.17}$$

and

$$\frac{\mathrm{d}y}{\mathrm{d}t} = -bx, \qquad b > 0; \tag{8.18}$$

they apply for $x > 0$ and $y > 0$. These form a coupled system and we cannot integrate either of the equations individually. The only equilibrium point is $x = 0$, $y = 0$, which represents mutual annihilation. Otherwise we note that, while x and y are still greater than zero, both are decreasing and, as the numbers fall, so do the rates of decrease diminish.

Eventually, we must reach a point where one of the armies has no men left and in our model this signals the end of the battle.

We can eliminate time from our equations simply by dividing. We get

$$\frac{\mathrm{d}y}{\mathrm{d}x} = \frac{bx}{ay}$$

or

$$ay\,dy = bx\,dx.$$

The variables are separated and we can integrate both sides to give

$$\frac{ay^2}{2} = \frac{bx^2}{2} + \frac{c}{2}$$

or

$$ay^2 = bx^2 + c.$$

If the starting conditions are $x = x_0$, $y = y_0$, then

$$ay_0^2 = bx_0^2 + c.$$

So our solution can be written

$$a(y^2 - y_0^2) = b(x^2 - x_0^2). \qquad (8.19)$$

In the x–y plane the solution curve is part of a hyperbola as shown in Fig. 8.15.

Suppose, as a special case, that the initial sizes of the two armies are such that $ay_0^2 = bx_0^2$; then equation (8.19) reduces to $ay^2 = bx^2$ or $\sqrt{a}y = \sqrt{b}x$. This means that the solution curve is the straight line through the origin with slope $\sqrt{b}/\sqrt{a}$ and the battle progresses inexorably from the point

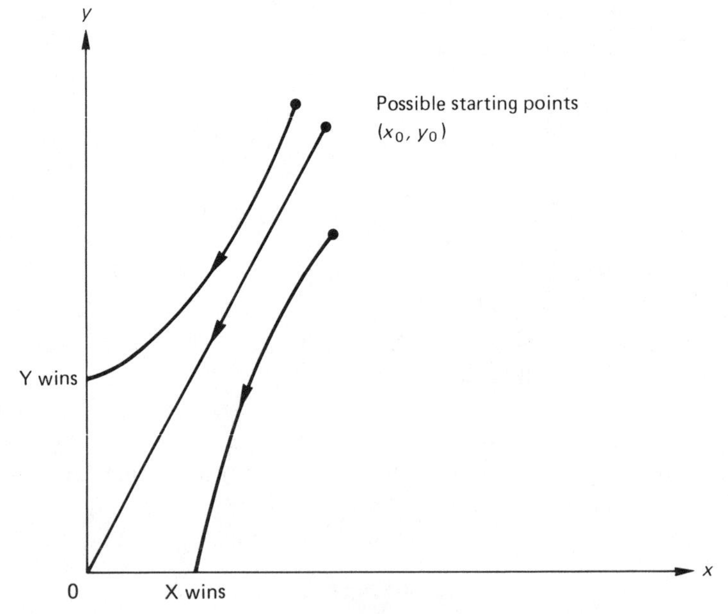

Fig. 8.15

(x_0, y_0) to the point $(0, 0)$. We conclude from this that the condition for an evenly matched battle (one that results in mutually assured destruction) is $ay_0^2 = bx_0^2$.

It follows that a sensible measure of any army's fighting strength can be obtained from the square of its size multiplied by its kill rate. This was pointed out by Lanchester in 1914 and is known as Lanchester's square law.

If $ay_0^2 > bx_0^2$, then the point (x_0, y_0) is above the straight line of equal fighting strengths in Fig. 8.15 and the Y army will win. We can also check this from the general solution (8.19).

For the X army to win, we would have to have $y = 0$ while x is still greater than zero, and from equation (8.19) we would then have $-ay_0^2 = b(x^2 - x_0^2)$ so that $x^2 = (bx_0^2 - ay_0^2)/b$. However, we have assumed that $ay_0^2 > bx_0^2$; so this equation has no solution in real numbers. On the other hand, equation (8.19) predicts that $x = 0$ when $y = \sqrt{(ay_0^2 - bx_0^2)/a}$.

According to our assumption that the initial values satisfy $ay_0^2 > bx_0^2$, this gives a meaningful value for y which is in fact our model's prediction for the number of survivors of the Y army at their moment of victory.

Returning to our initial example, let us choose an hour as our unit of time and assume that 0.15 X soldiers are killed by each Y soldier in 1 h and that 0.1 Y soldiers are killed by each X soldier in 1 h. Our difference equations are $\Delta x = -0.15y\,\Delta t$ and $\Delta y = -0.1x\,\Delta t$, where $\Delta t = 1$ h. We shall get approximately the same solutions if we solve the differential equations

$$\frac{dx}{dt} = -0.15y,$$

$$\frac{dy}{dt} = -0.1x.$$

Our starting values are $x_0 = 10\,000$ and $y_0 = 5000$. The fighting strength of the X army is $0.1 \times (10\,000)^2 = 10 \times 10^6$ while the fighting strength of Y is $0.15 \times (5000)^2 = 3.75 \times 10^6$. Our model therefore predicts that the X army will win the battle with $\sqrt{(bx_0^2 - ay_0^2)/b} = \sqrt{(10 \times 10^6 - 3.75 \times 10^6)/0.1} \approx 7906$ troops remaining. For the Y army to win their initial number of troops must be such that $y_0^2 > bx_0^2/a$, i.e. $y_0^2 > 0.1 \times (10\,000)^2/0.15$ or $y_0 > 8165$.

With the original values of $x_0 = 10\,000$ and $y_0 = 5000$, how long will the battle last? This is not so easy to answer but we can get estimates without actually solving the equations.

Initially, we have $(dy/dt)_{t=0} = -0.1x_0 = -1000$; so at the start of the battle the Y troops are being killed at the rate of 1000 per hour. If this rate continued (in reality, it will not because the X army is also being reduced) the Y army would be totally wiped out in $5000/1000 = 5$ h. The battle must therefore last *at least* 5 h. We can also find an *upper bound* for the duration of the battle by considering the rate at which the Y troops are being killed towards the end of the battle. We know from our calculations that there will be about 7906 X troops alive at that time; so $(dy/dt)_{\text{end of battle}} \approx -0.1 \times$

$7906 = -790.6$. If this had the same constant value from the beginning of the battle, then we would have the simple model $y = -790.6t + 5000$ which predicts $y = 0$ at $t = 5000/7906 \approx 6.32$.

We now know that the battle will last between 5 h and 6.32 h and a midpoint estimate would be 5.7 h. These thoughts are illustrated in Fig. 8.16. The exact answer can be found by solving the differential equations. Differentiating the first equation, we have

$$\frac{d^2x}{dt^2} = -0.15\frac{dy}{dt}$$

and substituting for dy/dt from the second equation, this becomes

$$\frac{d^2x}{dt^2} = 0.15 \times 0.1x$$

or

$$\frac{d^2x}{dt^2} - 0.015x = 0.$$

This has a solution of the form

$$x = A\exp(\sqrt{0.015}t) + B\exp(-\sqrt{0.015}t), \qquad (8.20)$$

where the constants A and B can be found from the facts that $x_0 = 10\,000$ and $y_0 = 5000$. Substituting $t = 0$, we have

$$10\,000 = x_0 = A + B. \qquad (8.21)$$

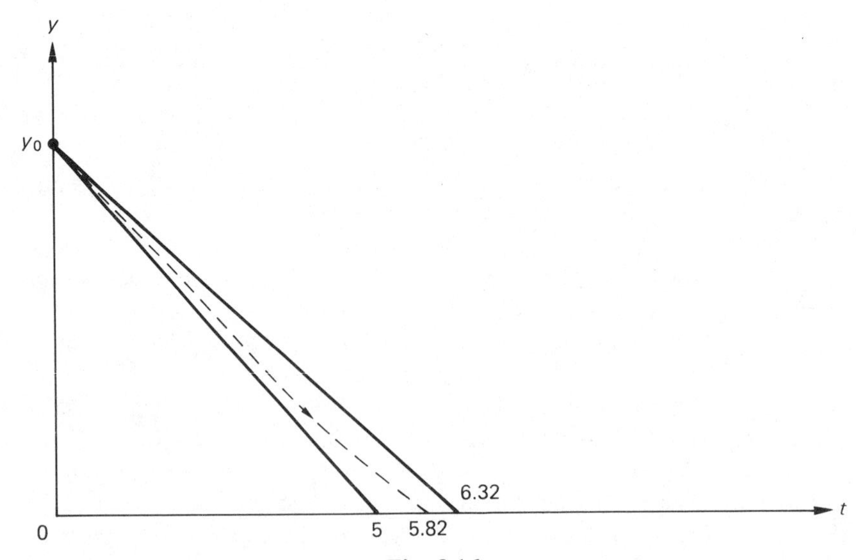

Fig. 8.16

The second fact is less easy to use. If we differentiate equation (8.19), we get

$$\frac{\mathrm{d}x}{\mathrm{d}t} = \sqrt{0.015}\,[A\exp(\sqrt{0.015}t) - B\exp(-\sqrt{0.015}t)]$$

and we know that

$$y_0 = \frac{\dot{x}_0}{-0.15}.$$

So

$$5000 = y_0 = \frac{\sqrt{0.015}(B-A)}{0.15}. \tag{8.22}$$

Solving equations (8.21) and (8.22) for A and B, we get $A \approx 1938.14$ and $B \approx 8061.86$. The size of the Y army at any time is

$$y = \frac{-\dot{x}}{0.15}$$

$$= \frac{-\sqrt{0.015}\,[A\exp(\sqrt{0.015}t) - B\exp(-\sqrt{0.015}t)]}{0.15}.$$

From this we deduce that $y=0$ when $\exp(2t\sqrt{0.015}) = B/A \approx 4.16$, which gives $t \approx 5.82$ h, confirming our previous estimate.

We calculated earlier that, for the Y army to win the battle, they would have needed 3165 extra troops at the beginning. Suppose reinforcements for the Y army become instantly available (e.g. paratroops) at some time after the beginning of the battle. We can calculate the number N needed for Y to go on to win the battle from the fact that we need at least to make up the difference in fighting strength at that moment, i.e.

$$N \geqslant \sqrt{\frac{b}{a}}x - y$$

$$= \sqrt{\frac{0.1}{0.15}}x - y.$$

This gives the following results.

t/h	0	1.0	2.0	3.0	4.0	5.0
N	3165	3577	4043	4570	5166	5839

Note that equations (8.17) and (8.18) constitute *a* simple battle model. Many other (and better) models are possible using different differential equations based on more careful assumptions.

Example 8.5.2: Foxes and rabbits

Problem statement

Suppose that some rabbits and foxes are living in a confined area where there is plenty of food for the rabbits and the foxes depend on eating rabbits for their food.

Objective

Develop a model which will enable us to predict the numbers of rabbits and foxes alive at any time t.

Formulate a mathematical model

Our variables are x, the number of rabbits at time t, and y, the number of foxes at time t. To proceed with the model, we need some assumptions. Suppose that left to themselves, i.e. in the absence of foxes, the rabbits would increase by 400% per time unit. If the time unit is very small, we can model this growth by the differential equation

$$\frac{dx}{dt} = 4x \qquad \text{when } y = 0.$$

If there were no rabbits, the foxes would starve to death. Suppose this would be at the rate of 90% per time unit. This means that (approximately) $dy/dt = -0.9y$ when $x = 0$. When both rabbits and foxes exist together, it is reasonable to assume that the number of rabbits killed in one time unit by foxes is proportional to the number of rabbits and also proportional to the number of foxes. Also, the more food there is, the healthier the foxes become and the more young foxes are born. With these interactions included, our differential equation for the rabbits takes the form

$$\frac{dx}{dt} = 4x - 0.02xy.$$

The constant 0.02 represents the proportion of the rabbit population killed by one fox in one time unit. (The value 0.02 is an arbitrary choice in the absence of any real data.) Similarly, our differential equation for the foxes becomes extended to

$$\frac{dy}{dt} = 0.001xy - 0.9y.$$

Here the constant 0.001 represents the proportion of extra births in the fox population per time unit due to the food value of one rabbit (again an arbitrary numerical choice).

Comparing the two differential equations that we have just written down with equations (8.17) and (8.18), we see similarities and differences. The xy terms on the right-hand sides of the equations, representing the interaction between the two populations, are again present but with a *positive* sign in the case of the foxes. Also both equations have a linear term representing a natural birth rate in the case of the rabbits and a natural death rate in the case of the foxes.

To investigate equilibrium points we set $dx/dt = 0 = dy/dt$, i.e.

$$4x - 0.02xy = 0 = 0.001xy - 0.9y.$$

One solution is obviously $x = 0$, $y = 0$, when there is no life at all. Another is given by solving $4 - 0.02y = 0$ and $0.001x - 0.9 = 0$, which gives $x = 900$ and $y = 200$. According to this model and under these conditions, 900 rabbits and 200 foxes would live 'harmoniously' together and their numbers would remain at these levels indefinitely. If we try the values $x(0) = 800$ and $y(0) = 100$, we find a cyclic behaviour in both rabbit and fox populations, as shown in Fig. 8.18. The rabbit population oscillates between about 168 and about 2658, the whole cycle repeating itself with a period of about 3.5 time units. The oscillation in the fox population is from just under 100 to about 350 with the same period.

In the phase diagram (Fig. 8.17), we see the rabbit and fox population levels plotted one against the other and the trajectory is a closed curve which is traversed once every 3.5 time units approximately. The initial point is

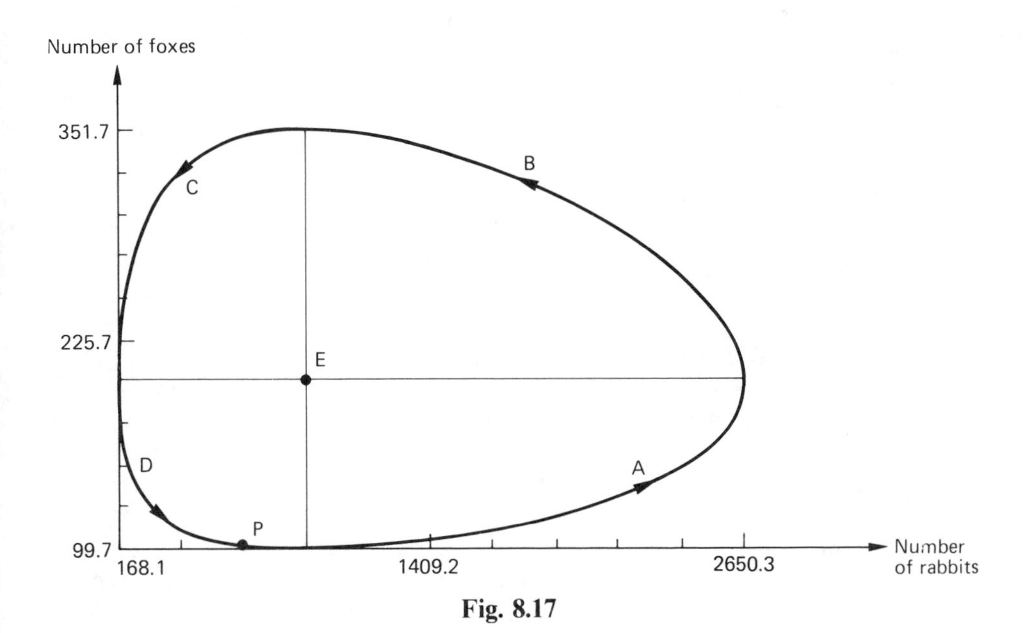

Fig. 8.17

marked P and the arrows show the direction in which the trajectory is traversed. The point E is the equilibrium point (900, 200).

At points along the section of the curve marked A, both populations are increasing. Note that in this region we have $x > 900$ and $y < 200$ and the differential equations give $\mathrm{d}x/\mathrm{d}t > 0$ and $\mathrm{d}y/\mathrm{d}t > 0$, confirming what we have just said. In the section marked B, having crossed the horizontal line where $y = 200$, we have $\mathrm{d}x/\mathrm{d}t < 0$, while $\mathrm{d}y/\mathrm{d}t > 0$. Our interpretation of this is that there are now so many foxes that the rabbit population is decreasing. In section C there is not enough food for the foxes and so they decrease as well as the rabbits ($x < 900$ and $y > 200$ and so $\mathrm{d}x/\mathrm{d}t < 0$ and $\mathrm{d}y/\mathrm{d}t < 0$).

Along section D, there are so few foxes that the rabbits are able to begin to increase their numbers again and we return to point P. Note that the maximum and minimum values of y occur when $x = 900$, the equilibrium value, and $\mathrm{d}y/\mathrm{d}x = 0$, while the maximum and minimum values of x occur when $y = 200$, the equilibrium value, and $\mathrm{d}y/\mathrm{d}x = \infty$. We cannot find the actual maximum and minimum values, however, without solving the differential equations.

If we start with different values of $x(0)$ and $y(0)$, we find a similar behaviour with a trajectory in the form of another closed curve going round the equilibrium point. However, if we make things very unfavourable for the rabbits by lowering their birth rate to 200% per time unit and revising upwards our estimate of the effect of the foxes on the rabbit population, then the differential equation for the rabbits reads

$$\frac{\mathrm{d}x}{\mathrm{d}t} = 2x - 0.1xy.$$

With the other differential equation unchanged, and starting with 100 rabbits and 100 foxes, we find that after one time unit we are down to less than one rabbit and about 400 foxes. If, regardless of this disaster for the rabbits. we continue with our model, we find that the rabbit population reaches a minimum of about 0.44 after 1.8 time units, while the fox population reaches a minimum of 0.38 after 6.8 time units, after which time both populations are increasing together.

We now realise that for the sake of realism we should have stated at the outset that our model applies only when $x > 1$ and $y > 1$. The conclusion from our last run of the model is that the rabbits would all have been wiped out after about 1.8 time units and the foxes would then either have to find an alternative food supply or die themselves.

You recall that we chose the coefficients in the differential equations arbitrarily and Figs 8.17 and 8.18 show the results. Curves similar to these have in fact been obtained from real data. An example is the interaction between the Canadian lynx and hare.

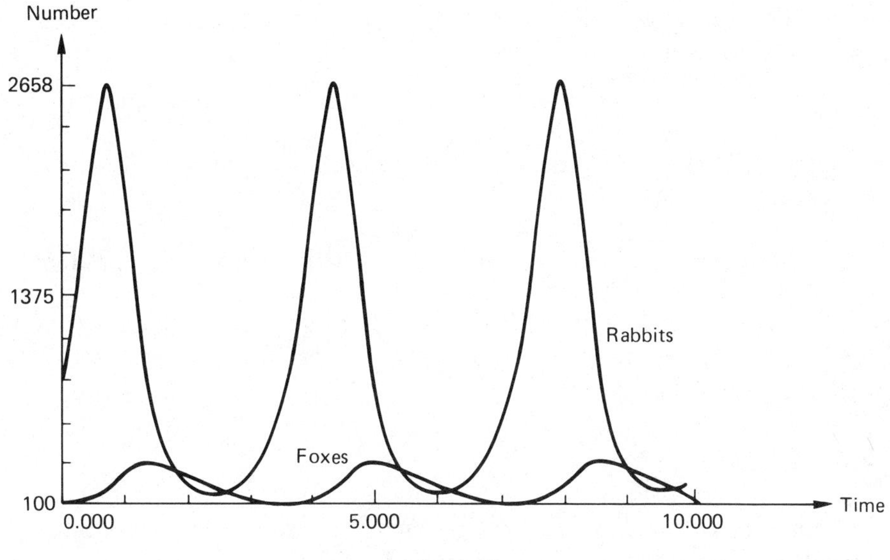

Fig. 8.18

EXERCISES 8.5

1 In the above battle model, if the Y army had the minimum necessary number of troops to win, how long would the battle last?

2 If instead of entering the battle *en masse* at a particular moment, the reinforcements arrived at a continuous constant rate, how would you modify the model?

3 Two different species of insect live together and eat the same food. They do not interfere with each other apart from the fact that they are competing for food. Make the following assumptions.

(a) For both species the number of births in a small time Δt is proportional to

(i) Δt,
(ii) the present population size and
(iii) the amount of food available.

(b) For both species the number of deaths in a small time Δt is proportional to

(i) Δt and
(ii) the present population size.

(c) The amount of food available is large but finite and steadily decreases as it is eaten by the insects.

Derive a mathematical model based on these assumptions and use it to investigate what happens to the two populations for various initial values and using different values of the coefficients in the differential equations.

4 Find what happens in the 'rabbits and foxes' model when different values are put for the coefficients in the differential equations. Can you extend the model to include grass? You will need to make assumptions about the rate of growth of grass as a function of time and the rate at which grass is eaten by the rabbits.

8.6 SUMMARY

In this chapter, we have looked at some models involving differential equations. This is a very important area of mathematical modelling since very many models involve variables changing with time. As with all modelling problems, there is no standard method of setting up the differential equations relevant to the particular problem in hand but the following points may be useful.

1 After writing down the problem statement and listing your assumptions carefully, read through and look for the words 'change', 'rate', 'increase', 'decrease', 'growth' and 'decay'. Any of these words will usually imply that a differential equation may be appropriate.
2 Differential equations result from applying an assumption or a 'law' governing rates of change. Is there a well-known scientific law which applies to your particular problem (such as Newton's mechanics) or do you have to derive one from your own assumptions? Very often the relevant differential equation will come out of the following very simple 'input–output' principle:

$$\begin{Bmatrix}\text{net rate}\\\text{of change}\end{Bmatrix}=\begin{Bmatrix}\text{rate of}\\\text{input}\end{Bmatrix}-\begin{Bmatrix}\text{rate of}\\\text{output}\end{Bmatrix}.$$

Although this seems very obvious, you will find in practice that, once you have written this down and substituted appropriate expressions involving the variables which appear in your model, you will have your differential equation in front of you.
3 Before rushing into trying to solve a differential equation, stop to think. Does the equation make sense in terms of what you know about how the model should behave? Check that each term in the equation has the same physical dimensions (see chapter 4). If time is the independent variable, the choice of time unit is up to you; it could be seconds, hours or years.

Choose whatever is most appropriate for the time scale of the model that you are dealing with. Make sure that any parameters, etc., that appear have values referred to the *same* time unit. Can you draw any conclusions from the equation without solving it? Are there any values of the variables for which the derivative is zero? What are the implications? Under what conditions are the derivatives positive, negative, very large and very small? Does the equation imply any restrictions on the values of the variables, e.g. the equation $\mathrm{d}y/\mathrm{d}t = \sqrt{x - 5}$ is meaningless if $x < 5$. Are there any restrictions on the variables arising from *physical* factors?

Solve the equation analytically if the integration looks easy. Remember that a single integration of a first-order differential equation will give you a *set* of solution curves with an arbitrary constant. Use any available particular information such as starting values of the variables to derive the values of arbitrary constants. If an analytic solution seems difficult to obtain, use numerical methods. There are many methods to choose from and also software packages for both large and small computers are available, especially designed for modelling problems involving differential equations.

9 REPORT WRITING AND PRESENTATIONS

9.1 INTRODUCTION

In the previous chapters, we hope that we have shown you that modelling is fun, exciting, challenging and rewarding. We may build models for enjoyment or to help us to make sense of the world around us but very often we build models in response to a request for help with a practical problem. Ultimately, we need to communicate our conclusions to the person or persons who asked for our help in the first place. These may be our employers, clients or friends and it is almost certain that they will not be mathematicians. They will only want to know what conclusions we have come to from our model and what advice we have to offer.

In this chapter, we focus our attention on how to communicate our conclusions and also how to explain the way in which we obtained our conclusions. This can obviously be done in two main ways: in writing and verbally. In practice, it is a good idea to use both. We shall refer to them as the (written) report and the (verbal) presentation. If we want our modelling efforts to be turned to good practical use we must make sure that our communication is *effective*. We might develop a very good and potentially very useful model but, if we do badly in communicating our conclusions, we may fail to convince anybody of its worth. With mathematical models, we need to take special care because we shall very often be communicating with people who have less mathematical training than ourselves.

9.2 REPORT WRITING

How much we write and how we say it will depend partly on how large the problem was and for whom the report is intended. We shall describe an 'average' report based on a straightforward problem. The guidelines which we describe should not be taken as absolutely rigid rules; some of the sections described can be shuffled around to suit a particular report.

The first point to make is that the report should have a well-defined *structure*. This will make it more likely that the intended message will get across to the reader and will make it easier to pick out parts of the report that may be of particular interest.

We can divide the structure into three sections, to be assembled together in the following order.

1 Preliminary.
2 Main body.
3 Appendices.

Each of these sections will be further divided into subsections of unequal lengths.

Preliminary

We start with the *front cover* which will carry the first important item of information—the title of the report. The title should be clear and direct, avoiding any technical jargon, and should be as brief as possible. It should contain the essence of what the report is about. The *title page*, which may be either the first page or the front cover itself, should contain the title, the names of the writers, their official capacity (if any), the organisation which they represent and the date of issue. The format will look similar to

A REPORT ON ....

by

....

....

Date: ....

On the next page an *acknowledgement* or thanks to persons or organisations which have helped in the investigation may be inserted. Following this should be the *contents*, listing the headings and subheadings of all the sections of the report and quoting the relevant page numbers. Appendices and illustrations should also be included in the list. As well as outlining what the report contains, the contents also communicate the structure of the report.

After the contents, there should be a *summary*. This constitutes a report in miniature covering all sections of the main report and it could be used as an 'abstract' for people who do not require the detailed information provided by the whole report. You should write the summary *after* the main report is finished and try to contain it within a single page.

If the reader is impressed by the contents and summary, he or she may be ready to read the main report but in some cases, e.g. a senior person with little time, he or she might consider it sufficient simply to ask for conclusions and recommendations. It may be a good idea therefore to include in the summary the main *conclusions*. (Full conclusions are given as the last section in the main part of the report.) These should sum up the results that have been obtained from the model and what the implications are. There may be recommended courses of action, offering advice based on the interpretation of the results produced by the model. In effect, we are providing a short-cut through the report, i.e. title–contents–summary and conclusions, as an alternative to reading the full report.

Before we leave the preliminary section, we have a further page to add, namely a glossary of variable names together with units of measurement and symbols used. Any particular variable or parameter should of course be represented by the same symbol throughout the report. Also included should be an explanation of any abbreviations used in the report.

Main body

This should start with a *problem statement* explaining the background to the problem and why the model was developed. The objectives of the investigation should be clearly stated here. (This subsection may alternatively appear in the preliminary section of the report.) You need to remember that whoever reads your report will look at the conclusions to see whether the objectives have been achieved; so you must ensure that they relate.

In developing your model, you will have made a number of simplifying *assumptions*. These should now be listed carefully and clearly. Include all the assumptions that you *had* to make in order to make progress with the model together with those assumptions that you *decided* to make in order to keep the model simple. There is obviously going to be a link between the assumptions and the final conclusions obtained and you could be challenged on the validity of the assumptions. You should be prepared to defend them if necessary.

It is very likely that some *data* are included in the report. Present the data clearly and accurately using tables and graphs and also give the *source*. If there is a large amount of data, this would be best put in the appendix.

The next subsection will probably be the longest in the whole report. It will explain the *development* of the model from the assumptions and the derivation of equations, etc., which completes the formulation. There may well be a number of diagrams involved also.

The development and formulation will be followed by the *solution of the equations*, including an explanation of the solution procedure and any computer software used (either specially written or standard packages). It may also be of relevance to mention the computer hardware used. If much algebraic manipulation and/or numerical work is involved, it may be better to postpone some of it to the appendices.

Having come to the end of the solution stage, we shall be ready to present our *results*. These could range from very brief (a single number or statement) to many pages of graphs or tables. If there *are* many pages of results, it is probably wiser to put them into the appendices. Do not produce a large volume of results if they are not really useful. The main point is that they should illustrate the conclusions and provide information that could be of use.

You may wish to sum up your *conclusions and recommendations* here instead of or as well as in the preliminary section. Do not forget to say something about the possible accuracy of your results and the sensitivity of the conclusions obtained to the parameter values used and the assumptions made. You may wish to criticise your own model here, pointing out its limitations.

Finally there could well be a section outlining possible *extensions*, generalisations and suggestions for further work.

Appendices

This is the section into which detailed information relevant to the subject of the report can be put. We may have given the impression that it is some kind of 'dustbin' into which we put things which are rather inconvenient. The point, however, is to enable an interested reader to study the information in greater depth than is possible in the main body of the report. It is also possible to include in an appendix items of information such as graphs or tables which are frequently referred to in the report and would otherwise have to be repeated several times.

In the main body of the report, you may have referred to, or directly quoted, or used data from published sources. These should be listed under a heading 'References', including author names, book or journal title and volume number, page numbers and date of publication. You may also know about books or articles not mentioned directly in the report but containing relevant material either as background reading or further reading. These should be listed under the heading 'Bibliography'.

Summary

Normally the following headings will appear in a written report but not necessarily in the same sequence. Your particular report may involve special factors which require a different sequence or you may decide to leave out some of the sections.

Front cover.

Preliminary.
Title page.
Acknowledgements.
Contents.
Summary.

Conclusions.
Glossary.

Main body.
Problem statement.
Assumptions.
Data.
Development.
Solution procedure.
Results.
Conclusions and/or recommendations.
Extensions.

Appendices.
Data.
Derivation of equations.
Program structure and/or flow diagram.
Program listing.
Numerical results.
Graphs, tables and illustrations.

References and/or bibliography.

Back cover.

General remarks

Before starting on the report, spend some time thinking. If you know who the readers are going to be, ask yourself 'What do they need to know?' and 'What do they know already?' Make rough notes of particular points that you want to make sure are in the report. Consider the purpose and scope of the investigation and aim for a report which is appropriate in length for the importance of the work that has been done. Also consider the time available for preparing the report and divide it up between thinking, planning, writing and revising. Decide on the style of writing, bearing in mind who your readers are going to be. If they are strangers to you, be appropriately formal but not too pompous or pedantic. Try to choose a style which is clear and informative, remembering that the object is to communicate what you have done.

While writing the report use A4 size paper and write or type on one side only, leaving a wide left-hand margin. If possible, use a word processor so that revisions can be easily and quickly made. Make sure that all pages are clearly numbered and always start new sections, subsections or topics on a new page. Give sections and subsections *headings* which are underlined. Also underline any points requiring special emphasis in the report. It is a good idea to use a numbering scheme such as

5 BODY OF REPORT

5.1 Problem statement

5.2 Assumptions

5.2.1 First assumption

etc.

The merit of a numbering scheme such as this is that it makes it easier to cross-refer from any part of the report. Do not go *too* far with it, however; for example a section such as 5.2.1.5 could well put off the reader.

Each graph or table should have its own page and be clearly titled. If possible, place them next to the section of text in which they are discussed. Be neat and well organised and make sure that the grammar and spelling are correct. If they are not, your reader will quickly be distracted even though you may have something good to say. Avoid long sentences (30 words or more) and do not use complicated words when simple words will communicate the meaning equally clearly. Also keep paragraphs reasonably short and well spaced. This makes a more effective visual impact than large blocks of print.

Consider *revising* your report after putting it on one side for a while. Read through your first draft and consider the following questions.

Is the report objective?
Is it likely to create a good impression?
Does the writing match the needs of your readers in style, vocabulary and level of mathematics?
Is the report positive and constructive?
Is the title page complete and well laid out?
Is the summary clear, brief and accurate?
Is the layout clear and easy to follow?
Does the report read well?
Are any essential sections missing?
Are the sections in the most suitable order?
Do the headings stand out?
Is the problem clearly stated?
Is the level of detail appropriate?
Have you included all the facts and no unnecessary information?
Are the sources of facts clear?
Do the conclusions follow logically from the development?
Are possible solutions abandoned without reason?
Are symbols, etc., suitable and consistent?
Are there any statements whose meanings are not quite clear?
Are facts, figures, calculations and drawings accurate?
Are the most important points sufficiently emphasised?
Are there any faults of logic or mistakes in spelling?

If you feel that the answers to any of these questions are not satisfactory, make the necessary revision until you are satisfied with what you have produced.

9.3 A SPECIMEN REPORT

The brief specimen report which follows describes a genuine attempt to solve a real practical problem. It is presented here as a specimen with which to illustrate *some* of the topics discussed in section 9.2. It should not be taken to represent an ideal or typical report. In practice, reports are certain to be much longer than this and to contain far more data and analysis of results.

A REPORT ON

TESTING FOR A RARE BLOOD CONDITION

by

D. Edwards and M. J. Hamson

Thames Polytechnic

January 1988

CONTENTS

1

1 PRELIMINARY SECTIONS

1.1 Summary and conclusions

The problem considered is how to identify all those individuals in a population who possess a certain rare blood condition, using the minimum number of tests. A two-stage procedure is found to be the most efficient in terms of minimising the expected total number of tests. In this procedure the population is divided into groups each containing K_1 individuals at the first stage and K_2 individuals at the second stage. The optimum values of K_1 and K_2 are independent of the total population size, depending only on the incidence of the condition being tested in the population.

1.2 Glossary

Symbol	Description
N	Number of individuals in a population
P	Proportion of individuals in the population who have a certain condition
K, K_1, K_2	Numbers of individuals in groups
L, L_1, L_2	Numbers of groups
X	Number of groups giving a positive response to the test
T	Total number of tests carried out

2 MAIN SECTIONS

2.1 Problem statement

Suppose that we have a finite population of N individuals, each with a fixed probability of having a certain (rare) condition and we wish to identify all the affected individuals. We assume that samples of blood are available from all the individuals concerned and that a test is available which will detect the presence of the condition. If the testing procedure is expensive or difficult to carry out, we shall wish to minimise the total number of tests required to find all the affected individuals.

Suppose that it is possible to apply the test to the *pooled* blood sample from a number of individuals such that a positive result will be indicated if at least one of the individuals is affected. We can then save on the number of tests since a negative result will allow all individuals in the pool to be cleared.

There are a number of different testing procedures which could be used in these circumstances and the problem posed here is to find a procedure which will minimise the expectation of the total number of tests.

2.2 Assumptions

For the remainder of this report we shall assume the following.

1 The test in question can be applied to a pooled sample of blood made up from the samples taken from the individuals.
2 The test is 100% reliable in the sense that a negative response means that all individuals who contributed to the pooled sample must be clear and that a positive response will only be obtained when at least one individual in the pool is affected.
3 The amount of blood sampled from each individual is sufficient to be divided into a number of parts for subsequent testing.
4 Every individual involved has the same inherent probability P of being affected.

2.3 Individual testing

The most direct procedure is of course to apply the test to every individual. We show in the next section that this is in fact the best procedure if P exceeds about 0.3 but for smaller values of P the use of pooled samples gives a saving in the expected number of tests.

2.4 Single-stage procedure

In this procedure, we divide the population into a number of groups (L, say) and test a pooled sample from each group. Each group contains $K = N/L$ individuals and each test is a Bernoulli trial with probability $(1-P)^K$ of a negative response. If X is the number of groups giving a positive response, the distribution of X will be binomial (L, P'), where $P' = 1-(1-P)^K$.

For each group with a positive response, all the individuals in that group will be tested. This requires XK tests. The expected number of tests is therefore

$$\begin{aligned} E(T) &= L + KE(X) \\ &= L + N[1-(1-P)^K] \end{aligned}$$

or, in terms of K, the size of each group,

$$E(T) = N\left(\frac{1}{K} + 1 - (1-P)^K\right). \qquad (1)$$

K has to be an integer of course but, if we regard the expression on the

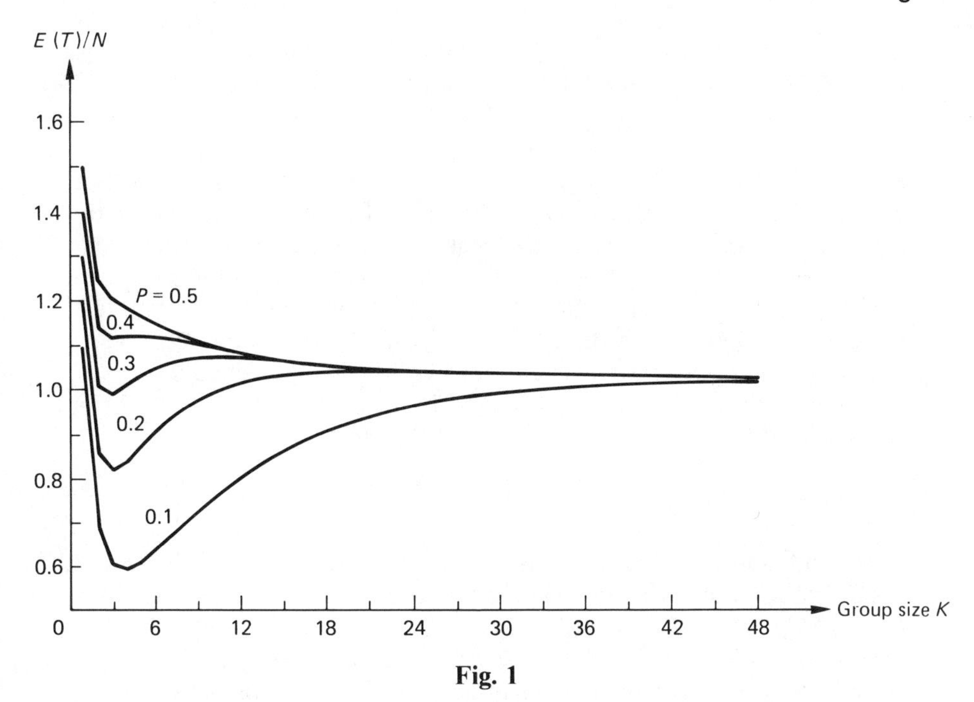

Fig. 1

right-hand side as a function of a continuous variable K, we obtain curves such as those shown in Fig. 1. For values of $P < 1 - \exp[-4\exp(-2)]$ (≈ 0.418), the curves have a local minimum for $K > 0$ so that the expected number of tests can be minimised by an appropriate choice of K.

Equation (1) shows that the optimal choice of K depends on P only. We can get an approximation to this optimal K value for small P by using the relation $1 - (1 - P)^K \approx KP$; then

$$E(T) = N\left(\frac{1}{K} + KP\right).$$

Differentiation with respect to K gives

$$\frac{d}{dk}\left(\frac{E(T)}{N}\right) = 0 \qquad \text{at } K = P^{-1/2}.$$

From the graphs in Fig. 1 we see that, as $K \to \infty$, we have $1/K + 1 - (1 - P)^K \to 1$ so that $E(T)/N$ can be brought close to 1 by choosing a sufficiently large value of K. In practice of course there is an upper limit on K given by the value of N. For values of $P \leqslant 1 - \exp[-\exp(-1)]$ (≈ 0.308), the local minimum for $E(T)/N$ does give the overall minimum. For values of P larger than this, however, the overall minimum for $E(T)/N$ is 1.0 and the best choice is to make K as large as possible, in other words to test each individual in the population.

The distribution of T/N can be found from the fact that $T/N = 1/K + KX/N$ where X is binomial $(N/K, 1-(1-P)^K)$.

2.5 Two-stage procedure

In this procedure the first stage is to divide the population into L_1 groups with $K_1 = N/L_1$ individuals in each group and to test the pooled blood samples from each. This requires L_1 tests to be carried out. Suppose that X_1 of these are positive. In the second stage, we divide all the positive groups into L_2 subgroups with $K_2 = N/L_1L_2$ individuals in each and test the pooled blood samples from each subgroup. This requires X_1L_2 tests. Suppose that X_2 of these tests are positive. For each positive result, we test every individual in that subgroup. The total number of tests for the whole procedure is

$$T = L_1 + X_1L_2 + \frac{X_2N}{L_1L_2},$$

where the random variable X_1 has the binomial distribution (L_1, P') and the distribution of the random variable X_2, conditional on X_1, is binomial (X_1L_2, P''), where $P' = 1-(1-P)^{N/L_1}$ and $P'' = 1-(1-P)^{N/L_1L_2}$. We have $E(X_1) = L_1P'$ and $E(X_2) = L_1L_2P'P''$, so that $E(T) = L_1 + L_1L_2P' + NP'P''$. As with the single-stage procedure, it is mathematically more convenient to express $E(T)$ in terms of K_1 and K_2, which gives us

$$E(T) = N\left[\frac{1}{K_1} + [1-(1-P)^{K_1}]\left(\frac{1}{K_2} + [1-(1-P)^{K_2}]\right)\right]. \qquad (2)$$

Using the approximation $1-(1-P)^x \approx XP$ for small P, we get

$$E(T) \approx N\left(\frac{1}{K_1} + \frac{PK_1}{K_2} + P^2K_1K_2\right).$$

On setting the partial derivatives with respect to K_1 and K_2 equal to zero, we find that for a stationary point of $E(T)$ regarded as a function of K_1 and K_2 we have

$$\left.\begin{aligned} K_1 &= \frac{1}{\sqrt{2}}P^{-3/4}, \\ K_2 &= P^{-1/2}. \end{aligned}\right\} \qquad (3)$$

In actual fact, of course, K_1/K_2 must be an integer. As with the single-stage procedure the optimal choices of K_1 and K_2 are independent of N, being dependent on P only.

2.6 Results

In Table 1 the values of K, K_1 and K_2 giving minimum $E(T)$ are recorded for various values of P. These were found by direct search from the exact

Table 1

P	Single-stage procedure		Two-stage procedure		
	K	$E(T)/N$	K_1	K_2	$E(T)/N$
0.001	32	0.063	50	25	0.0232
0.002	23	0.088	50	25	0.0285
0.003	19	0.108	48	16	0.0355
0.004	16	0.125	48	16	0.0426
0.005	15	0.139	42	14	0.0502
0.006	13	0.152	39	13	0.0575
0.007	12	0.164	36	12	0.0645
0.008	11	0.175	33	11	0.0712
0.009	11	0.186	30	10	0.0776
0.010	11	0.196	24	12	0.084
0.020	8	0.274	16	8	0.138
0.030	6	0.334	12	6	0.185
0.040	6	0.384	10	5	0.229
0.050	5	0.426	10	5	0.271
0.060	5	0.466	8	4	0.308
0.070	4	0.502	8	4	0.346
0.080	4	0.534	5	5	0.384
0.090	4	0.564	5	5	0.417
0.100	4	0.594	5	5	0.450
0.200	3	0.821	3	3	0.734
0.300	3	0.990	3	3	0.984
>0.308	N	1	N	0	1

expressions (1) and (2). A marked reduction in the expected number of tests is revealed when the two-stage method is used, especially at low values of P. The approximations given in equations (3) are found to be reasonably accurate. For example, at $P=0.02$, equations (3) give $K_1=13$ and $K_2=7$ while the actual optimum is at $K_1=16$, $K_2=8$. To indicate the distribution of T, a simulation was carried out using a random-number generator. The following results were obtained from 300 runs using a population size $N=96$ (for convenience) and $P=0.02$. The distribution of T using the two-stage procedure with $K_1=16$ and $K_2=8$ is as follows.

T	6	8	10	12	14	16	18	20	22	24	26	28
Frequency	41	89	42	32	13	20	15	14	3	5	11	5

6

T	34	38	40	42	48
Frequency	1	2	4	2	1

This has a mean value $\bar{T}$ of 12.76 and a variance of 57.42. The calculated value of $\bar{T}/N$ is 0.133, which agrees well with the theoretical prediction of 0.138 from Table 1.

For comparison, a simulation of 300 runs using the single-stage procedure with $K=8$ (for $N=96, P=0.02$) gave the following distribution of T.

T	12	20	28	36	44	52	60
Frequency	41	102	67	51	18	17	4

This has a mean value $\bar{T}$ of 27.2 and a variance of 128.64. The calculated value of $\bar{T}/N$ is 0.283 compared with the theoretical value of 0.274.

2.7 Regular section procedures

In a bisection procedure an initial test of the pooled sample for the population, if proved positive, is followed by a division of the population arbitrarily into two halves and a pooled test is applied to each half. Every positive result leads to another bisection until we get down to the level of individuals. Exact expressions for the number of tests required seem difficult to obtain. Simulation runs were carried out with $N=256$ for the bisection procedure and $N=243$ for a similar trisection procedure. Typical results are shown in Table 2 although the variance is appreciable with a coefficient of variation of about 2%.

The bisection results are better than the equivalent results for the single-stage procedure but less good than the two-stage results. The trisection results apparently show it to be the least efficient of the procedures considered.

2.8 Conclusions

Of the procedures examined, the most efficient in terms of minimising the expected total number of tests is the two-stage procedure in which the population is divided into groups each containing K_1 individuals at the first stage and K_2 individuals at the second stage. The optimal values of K_1 and K_2 are independent of the population size N, depending only on the value of P, the incidence of the condition being tested in the

Table 2

P	$E(T)/N$ for bisection	$E(T)/N$ for trisection
0.01	0.110	0.428
0.02	0.195	0.675
0.03	0.253	0.733
0.04	0.312	0.787
0.05	0.352	0.827
0.10	0.531	0.978
0.20	0.746	1.193
0.30	0.864	1.310
0.40	0.927	1.381

population. For small P the saving obtained over the single-stage test is about 50%. It is possible that an improvement could be made using procedures with more than two stages but, in practice, two stages seem likely to be sufficient.

3 APPENDICES

3.1 Possible extensions

1 We have assumed that all the individuals tested have the same probability P of possessing the condition. If instead it is known that age or sex or some other easily identifiable factor affects an individual's probability of having the condition, then this could be taken into account when forming the groups. A more efficient procedure might then be possible.

2 We have described the testing procedures in terms of blood samples but the same conclusions could apply to more general situations. Suppose that some finite population is considered in which every item either does or does not possess a certain defect. We assume that from each item a product can be extracted ('blood') on which a test can be carried out which will reveal the presence of this defect. We wish to identify all the defective items using the minimum number of tests. If the test can be applied to pooled samples as effectively as to individual samples, then the conditions are equivalent to those considered in this report and the same conclusions will apply.

3.2 Mathematical analysis

The graph of $E(T)/N$ against K for the single-stage procedure corresponds

8

to an equation of the form

$$y=\frac{1}{K}+1-x^K,$$

where

$$x=1-P.$$

To find the local extreme values of y, we differentiate with respect to K. This gives

$$\frac{dy}{dK}=-\frac{1}{K^2}-x^K \ln x$$

which equals 0 when

$$K^2x^K=\frac{-1}{\ln x}.$$

Now, since $x<1$, the expression K^2x^K considered as a function of K has a graph such as that sketched in Fig. 2.

There is a local (and global) maximum for some value $K=K_*$. We are looking for points where the graph is cut by the horizontal line at height $-1/\ln x$. Clearly there will be no such points if $-1/\ln x>m$, where m is the maximum value of K^2x^K, i.e. $K_*^2x^{K_*}$.

We find K_* by solving

$$\frac{d}{dK}(K^2x^K)=0$$

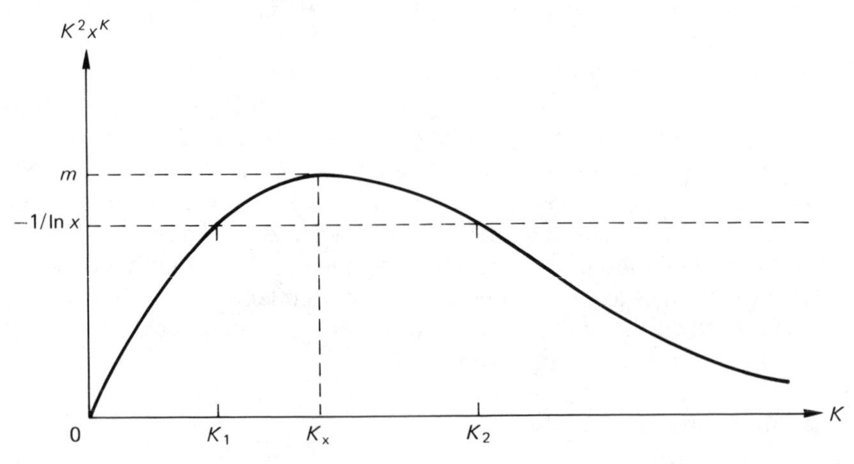

Fig. 2

or

$$\frac{\mathrm{d}}{\mathrm{d}K}(2\ln k + K\ln x) = 0,$$

i.e.

$$\frac{2}{K} + \ln x = 0.$$

Therefore,

$$K_* = -\frac{2}{\ln x}.$$

There will be no roots if

$$-\frac{1}{\ln x} > \left(-\frac{2}{\ln x}\right)^2 x^{-2/\ln x},$$

i.e.

$$x < \exp[-4\exp(-2)]$$

or

$$P > 1 - \exp[-4\exp(-2)]$$
$$\approx 0.418.$$

From the sketch, we see that, if we have two roots K_1 and K_2 and the graph of y against K behaves as shown in Fig. 3, K_1 is clearly the value

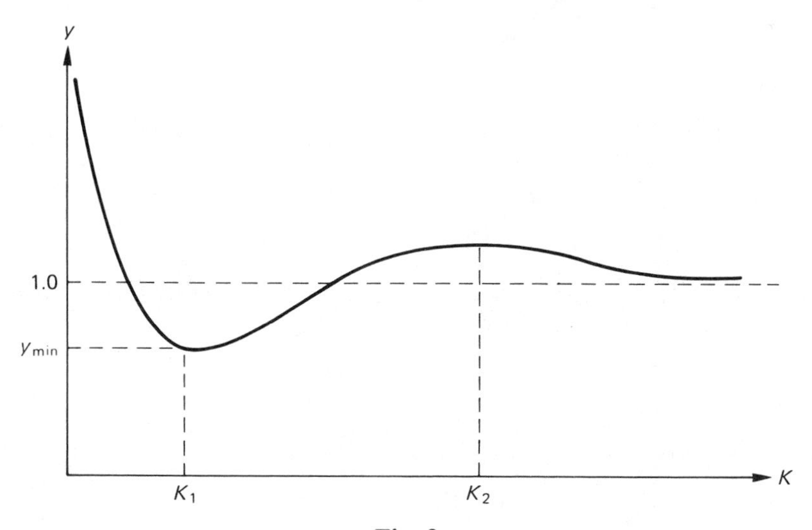

Fig. 3

10

in which we are interested. K_2 is in fact quite large. For example, when $P=0.1$, $K_2 \approx 55$.

The minimum value of y is

$$y_{\min} = \frac{1}{K_1} + 1 - x^{K_1}$$

and this will be less than 1 if $1/K_1 < x^{K_1}$. Combining this with the equation

$$\frac{1}{K_1^2} + x^{K_1} \ln x = 0,$$

we have that $y_{\min} < 1$ when

$$P < 1 - \exp[-\exp(-1)]$$
$$\approx 0.308.$$

9.4 PRESENTATION

Writing a good report, as we have emphasised in the previous section, is vital. In real life, time is often short, many different interests are competing for attention and there may be a reluctance to read reports (especially if they are lengthy). An alternative way to report your modelling conclusions to management and staff is to give a verbal presentation.

The precise arrangement will vary according to circumstances but, generally speaking, a time and place will be devoted to a verbal presentation of your findings in front of an audience (usually a small group). If you try to do this without adequate preparation, it will almost certainly be a disaster.

Preparation

You will normally have worked as a group of people on your modelling project and it is very likely that the presentation will involve all members of the group. Quite clearly this means it will be necessary

(a) to plan,
(b) to prepare and
(c) to rehearse your presentation.

For the *plan*, you need to consider the following questions. How much time will you have for the presentation? How much material should you try to present in that time (and *not* how much you can get through but how much the audience can take in)? In what order should you present the material? How should you share out the time between the members of your group? What are the most important points that you wish to make in your presentation? What will be the most effective way of making these points? What technical aids will be available to you for the presentation?

For the *preparation*, you need to keep in mind the order in which you will go through the material and where the 'natural breaks' will occur when one speaker hands over to the next member of your group. The most popular mechanical aid is the overhead projector (OHP). The material that you wish to present should be prepared on transparent sheets beforehand. Do *not* write or do mathematical calculations during the presentation itself. Try to arrange your material so that it comes naturally in sections of one transparent sheet at a time. Do *not* put too much information on one sheet and remember that your material should be legible and clearly visible from the back of the demonstration room. Data and drawings should be made *large* and clear. A superb drawing, if it is too small, is a waste of time. Use different coloured inks if you have them. As a rough guide, seven distinct pieces of information on one sheet should be the upper limit. Plan what you are going to say in advance and coordinate your speech with your OHP slides. Remember that your audience will be looking at the screen at the same time as listening to what you are saying. For the *rehearsal*, when you have got all the material ready, try to get the help of an impartial observer willing to sit and listen. *Time* your presentation carefully.

Giving the presentation

Have your material ready with all your OHP slides numbered and in correct sequence. Be punctual and arrive neatly dressed (impressions matter). Ideally the OHP equipment, etc., should already be set up with chairs arranged for both your audience and the members of your group (arranged at the front, facing the audience). If necessary, check up on the arrangements beforehand and make sure there will be no last-minute hitches such as being unable to get the OHP to work! A very useful idea is to supplement the OHP with a board on which is pinned a large piece of paper 'sign-posting' your presentation. This should show the headings for the various sections and a useful device is to have a movable 'arrow' which can be moved down the sheet as the presentation progresses. The audience will then be fully in the picture at all times, knowing where they are and what is to come next.

The *introduction* is a very important part of your presentation. This is where you make your opening impact and you must capture your audience's attention and interest. If you get this wrong, the rest of the presentation may flop. The responsibility of doing the introduction should be given to the most able member of your group. He or she will have the responsibility of setting the style for the whole presentation. One of the opening remarks should be to introduce all the members of the group by name, possibly indicating their particular roles in the presentation.

The *hand-over* from one speaker to the next is very difficult to do well and should be practised in the rehearsals. It should be smooth and efficient; otherwise the continuity of the presentation may be destroyed and the audience's concentration may wander. Some phrase such as 'I now hand you over to my colleague X who will...' should be used.

While *speaking*, do *not* read from notes if at all possible. This is guaranteed to put off the audience. Try to memorise what you are going to say. If you need notes, make them very brief and put them on cards rather than on a sheet of paper. If you are at all nervous, your hand may start to shake and this will be amplified if you are holding a large piece of paper! Look at the audience, make eye contact and look for responses. A nod or a smile shows that you are getting through. If they appear puzzled, slow down or repeat what you have just said or offer to answer any questions. Do not just carry on like a steamroller.

Try to avoid mannerisms; these are due to nervousness and can be controlled. Take care not to obscure the audience's view of the OHP screen. Do not turn round to look at it yourself; if you need to, read it off the machine itself. Speak clearly without rushing and never give the impression that you cannot wait to finish and sit down again.

Be prepared to answer questions from the audience at any moment. Answer politely even if you think it is a stupid question. If you do not know the answer, do not panic and do not try to avoid the question or fudge the issue! Say clearly that you do not know the answer and offer to look into it.

The *closing statement* is also very important and should be thought out carefully at the planning stage. Try to finish on a positive constructive note. At all costs, avoid finishing by trailing off into an embarrassing silence. It is a good idea for the first speaker to come on again for the finish; this rounds the presentation off nicely.

10 EXAMPLE MODELS

10.1 INTRODUCTION

In this chapter we look at a selection of models from a variety of backgrounds. There is no common theme and the models are not developed in detail. The aim of this chapter is to illustrate and complement the work of the previous chapters and it gives us an opportunity to put into practice some of the principles discussed.

As far as is practical, we have followed the methodology of chapter 3 in order to emphasise again that the underlying modelling process involves the same stages even when the individual problems vary widely in context. You should adopt this practice, or one similar to it, in your own modelling efforts with the qualification that the methodology is to be regarded as a helpful framework rather than a compulsory strait-jacket. For reference an outline of the methodology is repeated here.

Context.
Problem statement, objective, given . . . , find . . . .
Formulate a mathematical model, list factors and assumptions.
Obtain the mathematical solution.
Interpret the mathematical solution, validate the model.
Using the model, further thoughts.

The examples given in this chapter are not all complete; in fact, there are many questions left unanswered. You should read each modelling development critically. Try out your own ideas on these models and improve on them if you can.

10.2 DOING THE DISHES

Context

Consider a domestic scene; you have an enormous pile of greasy plates to wash and a bowl of hot soapy water. The water is hot enough to wash the grease off the plates but also cool enough for you to insert your hands. As the washing-up proceeds, the water gradually cools until it eventually becomes too cool to clean the plates properly.

Problem statement

How many plates can be washed using one bowl of hot washing-up water? *Given* the relevant physical data, *find* the number of plates.

Formulate a mathematical model

This problem involves plates, water and air; so we can use these as 'factor headings'.

Factors concerning the water.
Amount.
Initial temperature.
Surface area.
Final temperature.
Water flow.
Specific heat capacity.
Heat transfer coefficient.

Factors concerning the plates.
Number.
Size.
Initial temperature.
Final temperature.
Specific heat capacity.

Factors concerning the air.
Temperature.
Convection currents.

(We are leaving the bowl itself out of the model on the assumption that there is negligible heat exchange between the water and the bowl and between the sides of the bowl and the air.)

We now list our variables and parameters as in Table 10.1.

The data are as follows: $C_p = 600\ \mathrm{J\ m^{-3}\ K^{-1}}$ (ceramic); $C_w = 4200\ \mathrm{J\ m^{-3}\ K^{-1}}$; $M_p = 0.5$ kg; $M_w = 15$ kg; $T_a = 20\ °\mathrm{C}$; $T_w(0) = 60\ °\mathrm{C}$; $A = 0.1\ \mathrm{m}^2$; $T_f = 40\ °\mathrm{C}$; $h = 100\ \mathrm{W\ m^{-2}\ K^{-1}}$. Note that we have taken particular values of the parameters in order to calculate a specific answer.

Assumptions

1 We have already assumed that the bowl itself is not involved in any heat exchange.

Table 10.1

Description	Type	Symbol	Units
Number of plates	Output variable	n	Integer
Mass of one plate	Input parameter	M	kg
Air temperature	Input parameter	T_a	K
Water temperature	Output variable	T_w	K
Initial water temperature	Input parameter	$T_w(0)$	K
Final water temperature	Input parameter	T_f	K
Mass of water	Input parameter	M_w	kg
Water surface area	Input parameter	A	m^2
Heat transfer coefficient from water to air	Input parameter	h	$W\,m^{-2}\,K^{-1}$
Specific heat capacity of plates	Input parameter	C_p	$J\,m^{-3}\,K^{-1}$
Specific heat capacity of water	Input parameter	C_w	$J\,m^{-3}\,K^{-1}$

2 Assume that we wash one plate at a time by inserting it into the water, leaving it there for some time ΔT (possibly doing some scrubbing) and then taking it out to drain.
3 Assume that the amount of water in the bowl will remain constant, although in reality some may be lost owing to splashing and some may be taken away with the clean plates as we take them out. (There may also be a very small loss due to evaporation.)
4 Assume that the plates are initially at the air temperature T_a.
5 Assume that ΔT is long enough for the plate to reach the same temperature as the water (and also long enough to get the plate clean!).
6 Assume that ΔT is the same for all the plates (clearly not quite compatible with assumption 5 since it will take longer to get each plate clean as the water gets cooler).
7 Assume that the water loses heat by radiation and convection from the surface as well as by conduction to the plates and in melting the grease off the plates.

The principle that we shall use for developing our model (Fig. 10.1) is that heat energy is conserved. The quantity of heat energy contained within a body of mass M is McT, where c is a property of the material from which the body is made (called the specific heat capacity) and T is the temperature of the body in kelvins.

Here T'_w is the water temperature at the end of the interval, and E_c and E_r are the energy losses by convection and radiation, respectively. Heat energy balance gives us the equation

$$M_p C_p T_a + M_w C_w T_w = M_p C_p T'_w + M_w C_w T'_w + E_c + E_r. \qquad (10.1)$$

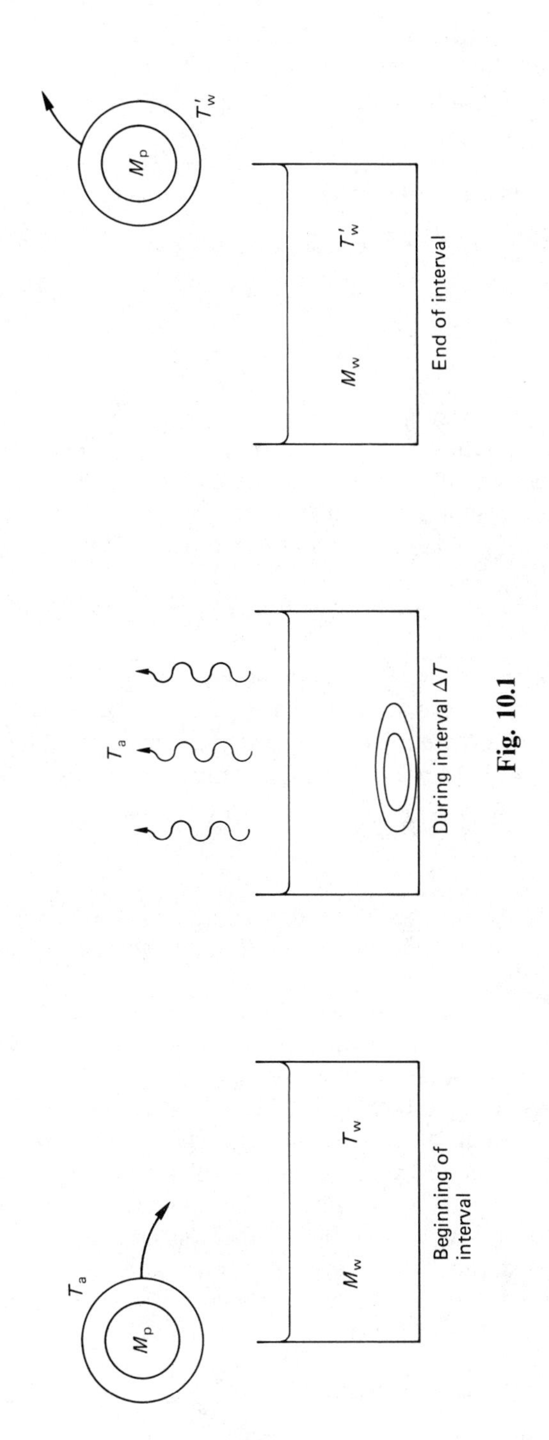

M_p
T_a
M_w
T_w
Beginning of interval
T_a
During interval ΔT
M_p
T'_w
M_w
T'_w
End of interval

Fig. 10.1

Let us forget the plates for a moment and consider the water left to cool by itself. There are well-established models for E_c and E_r as follows:

$$\text{rate of heat energy loss by convection} = hA(T_w - T_a), \qquad (10.2)$$

where h is the heat transfer coefficient and A is the surface area.

$$\text{rate of heat energy loss by radiation} = \varepsilon\sigma A(T_w{}^4 - T_a{}^4), \qquad (10.3)$$

where σ ($\approx 5.67 \times 10^{-8}$ W m^{-2} K^{-4}) is the Stefan–Boltzmann constant and ε measures the emissivity of a surface ($\varepsilon = 1$ for a black body and $\varepsilon \approx 0.95$ for water).

Note that these refer to instantaneous *rates* of heat loss and, before we go further, we must decide whether to represent our main variable in this problem, T_w, as a continuous variable or as a discrete variable. This is an important decision because the two choices lead us on quite different paths. If we adopt a continuous representation for $T_w(t)$, then, for water left to cool by itself, equations (10.2) and (10.3) translate into the *differential* equation

$$M_w C_w \frac{dT_w}{dt} = -hA(T_w - T_a) - \varepsilon\sigma A(T_w{}^4 - T_a{}^4), \qquad (10.4)$$

which has to be solved for $T_w(t)$, where t is the time measured from the initial state $t = 0$. In the discrete version, we represent T_w as the discrete function $T_w(n)$, where n represents the number of time steps Δt since the initial state. In this case equations (10.2) and (10.3) translate into the *difference* equation

$$M_w C_w[T_w(n+1) - T_w(n)] = -hA[T_w(n) - T_a]\Delta t - \varepsilon\sigma[T_w{}^4(n) - T_a{}^4]\Delta t. \qquad (10.5)$$

Different versions of this can be written, depending on whether we replace T_w by $T_w(n)$, $T_w(n+1)$ or some combination of the two (such as the mean $0.5[T_w(n) + T_w(n+1)]$).

This problem clearly has a natural 'discreteness' about it; so the best choice is probably the discrete representation and we can use ΔT, the time of washing one plate, as the basic time step Δt.

If we think of the process of washing one plate as a sudden immersion of a cold plate into hot water when the plate's temperature jumps instantly from T_a to T_w, then, for the rest of the time interval ΔT, equation (10.4) or its equivalent, equation (10.5), will apply. This means that we do not have to model the rate of heat transfer from the water to the plate during the time ΔT. (This *could* be done; the transfer mechanisms would be conduction and convection and we would need to know the thermal conductivity of the plate and the heat transfer coefficient for convection in water.)

Obtain the mathematical solution

It looks as if we are going to have rather a difficult equation to solve; can we simplify it? It is the T^4 term, coming from the radiation losses, which makes it difficult. Remembering our comments about relative sizes of terms (see section 5.7), we shall try to compare the numerical values of the radiation and convection losses. For this, we need data. The values quoted earlier were partly guesswork but they seem fairly realistic. We calculate

$$\begin{aligned} E_c &= hA(T_w - T_a) \\ &\approx 100 \times 0.1(333 - 293) \\ &\approx 400 \text{ J} \end{aligned}$$

and

$$\begin{aligned} E_r &= \varepsilon\sigma A(T_w{}^4 - T_a{}^4) \\ &\approx 0.95 \times 5.67 \times 10^{-8} \times 0.1[(333)^4 - (293)^4] \\ &\approx 26.54 \text{ J.} \end{aligned}$$

So the radiation loss is only about one-fifteenth of the convection loss and we could claim fair justification for dropping the E_r term.

Substituting $E_c = hA[T_w(n) - T_a]$, $E_r = 0$ and $T_w' = T_w(n+1)$ into equation (10.1) and rearranging, we have

$$(M_wC_w + M_pC_p)T_w(n+1) = (M_wC_w - hA\,\Delta T)T_w(n) + (M_pC_p + hA\,\Delta T)T_a.$$

Putting in the data values and taking $\Delta T = 10$ s (is this reasonable?), we get

$$\begin{aligned} (15 \times 4200 + 0.5 \times 600)T_w(n+1) = {} & (15 \times 4200 - 100 \times 0.1 \times 10)T_w(n) \\ & + (0.5 \times 600 + 100 \times 0.1 \times 10)293 \end{aligned}$$

or

$$T_w(n+1) = 0.9937T_w(n) + 1.8515.$$

This has the solution

$$T_w(n) = B(0.9937)^n + \frac{1.8515}{1 - 0.9937},$$

where $333 = T_w(0) = B + 1.8515/0.0063$. Therefore $B \approx 40$. So our model is

$$T_w(n) = 40 \times (0.9937)^n + 293.89.$$

We want the value of n when $T_w(n)$ equals the final temperature $T_f = 313$ K; so we find n from

$$313 = 40 \times (0.9937)^n + 293.89$$

or

$$0.4778 = (0.9937)^n.$$

Taking logarithms, $n = \ln(0.4778)/\ln(0.9937) \approx 117$.

Interpret the mathematical solution

Our model predicts that 117 plates could be washed under these conditions before changing the water. To *validate* the model, we would need a thermometer for the air and water temperatures and a mass balance to determine the masses of the plates and the water. We would also need to measure the dimensions of the bowl so that we can work out the surface area of the water.

Using the model

1 How does n vary with the other parameters? Plot graphs of n against

(a) M_w,
(b) M_p,
(c) $T_w(0)$,
(d) T_f and
(e) ΔT,

where in each case all the other parameters are held constant.

2 How would you modify the model if we put in more than one plate at a time but take them out one at a time?

3 Assumption 5 could be challenged. A possible replacement would be to assume that a constant amount of *energy* is required to clean each plate. Develop the model using this assumption.

10.3 SHOPPING TRIPS

Context

Where do people do most of their shopping? If we look at shopping trips made by the residential population of an area over a period of time, can we develop a model which fits the behaviour? Such a model could be used to predict the consequences of proposed redevelopments or new transport policies.

Problem statement

Given physical and economic data relating to shopping expenditure, *find* the shopping pattern.

Suppose that we divide a region into a number of 'zones', each of which has its residential population as well as shopping facilities.

Table 10.2 gives the distance in miles between the centres of three zones of a region and also the average shopping trip distances within each zone.

We are avoiding the question of what exactly we mean by a 'zone' and its 'centre'. In a real situation, there will obviously be a number of options and we would be guided by geographical and economic factors.

Table 10.2

	Distance/miles		
	To A	To B	To C
From A	2	7	5
From B	7	3	4
From C	5	4	3

At present the resident population figures are

A, 5800; B, 9400; C, 10 600;

the number of shops in each zone are

A, 22; B, 80; C, 220.

Formulate a mathematical model

Some of the factors involved are population, distances, number of shops, types of shop, transport facilities, travelling costs and shopping expenditure.

What causes people to choose to shop in a particular zone? Attractions are the number of shops and ease of access. Disincentives are distances, travelling costs and travelling time.

The factors involved in formulating a model are given in Table 10.3.

Our objective is to formulate a model relating x_{ij} (the number of shopping trips made from zone i to zone j in a specified period of time) with N_j, d_{ij} and P_i. Alternatively, we could have defined x_{ij} in terms of the amount of money spent on shopping during that period.

Table 10.3

Description	Type	Symbol	Units
Populations of the three zones	Input parameters	P_1, P_2, P_3	Integer
Distance between zone i and zone j	Input parameters	d_{ij} $(i=1, 2, 3; j=1, 2, 3)$	Miles
Number of shops in zone i	Input parameters	N_i $(i=1, 2, 3)$	Integer
Number of shopping trips by people of zone i in zone j	Variables	x_{ij} $(i=1, 2, 3; j=1, 2, 3)$	Integer

Assumptions

1 x_{ij} is proportional to P_i. (The more people live in zone i, the more trips there will be from zone i.)
2 x_{ij} is proportional to N_j. (The more shops there are in zone j, the more shoppers will be attracted there.)
3 x_{ij} decreases as d_{ij} increases. (Distance will put people off.)

The mathematical consequences of assumptions 1 and 2 are easy to formulate but what form could we use for assumption 3? Some possibilities are as follows.

(a) $x_{ij} \propto -d_{ij}$.

(b) $x_{ij} \propto \frac{1}{d_{ij}}$.

(c) $x_{ij} \propto \frac{1}{d_{ij}^2}$.

(d) $x_{ij} \propto \frac{1}{d_{ij}^\alpha}$ where α is a parameter.

(e) $x_{ij} \propto \exp(-\beta d_{ij})$ where β is a parameter.

Models such as this have been developed and used successfully (see, for example, D. Foot, *Operational Urban Models*, Methuen, 1981).

Generally, it is found that the effect of distance is quite pronounced and that x_{ij} decreases with d_{ij} more sharply than both the linear model (a) and the inverse model (b). The inverse square model (c) seems to be about right and by analogy with the inverse square law of Newton's mechanics this is often known as the gravity model. Many researchers have constructed useful models using the more general forms (d) and (e). These have the advantage of a parameter so that the model can be calibrated to some particular application. Usually, it is found that a value of α between 1.5 and 3 gives the best fit.

In this example, we shall use the inverse square model (c). Combining this with assumptions 1 and 2, we have

$$x_{ij} \propto \frac{P_i N_j}{d_{ij}^2}.$$

To change the proportionality into an equation, we must put in some constant k so that $x_{ij} = kP_i N_j / d_{ij}^2$.

Obtain the mathematical solution

The most practical way to proceed with the model is to calculate the quantities $P_i N_j / d_{ij}^2$ for all i, j and to find the sum $S = \sum_i \sum_j P_i N_j / d_{ij}^2$. Remembering that our objective was to predict the shopping *pattern*, we can do this in terms of percentages by dividing all the P_iN_j/d_{ij}^2 by S. This is done for the current example in Tables 10.4 and 10.5.

The values P_iN_j/d_{ij}^2 are as tabulated and $S=\sum_i\sum_j P_iN_j/d_{ij}^2=630\,874$. Dividing by S, we find the ratios given in Table 10.5. These can be converted into actual numbers of shopping trips by multiplying by the constant k. The shopping *pattern* however can be seen in Table 10.5 without using the k value.

Table 10.4

	P_iN_j/d_{ij}^2		
P_i	$N_j=22$	$N_j=80$	$N_j=220$
5 800	31 900	9 469	51 040
9 400	4 220	83 556	129 250
10 600	9 328	53 000	259 111

Table 10.5

	Shopping trips (proportion)		
	To A	To B	To C
From A	0.0506	0.0150	0.0809
From B	0.0067	0.1324	0.2049
From C	0.0148	0.0840	0.4107
Totals	0.0721	0.2314	0.6965

Interpret the mathematical solution

Our model predicts that, in the area covered by our model, about 7% of the shopping will be done in zone A, 23% in zone B and 70% in zone C.

We could be more flexible than this by allowing the 'constant' k to be different for different parts of the table. For example, if the three zones differed significantly in average per capita income, we would expect the more affluent zones to spend more on shopping. We could use different k values, k_1, k_2 and k_3, for the three lines in the table to reflect different 'spending powers'. To fit the model, we would need data on average incomes.

Note that, instead of distances between the zones, we could substitute travel times or costs. Instead of the number of shops as the attraction, we could substitute the number of *different* shops, the size of the shops (measured by floor space), the amount of car parking space, or some combination of these.

Using the model

1 A proposed new shopping centre in zone A will mean an additional 20 shops in that zone. What effect will this have on the shopping pattern?
2 A proposed new motorway will shorten the effective distance between zones A and B to 5 miles. How will this affect the shopping pattern?

10.4 DISK PRESSING

Context

You are employed to give advice to the production manager of a manufacturing firm. Part of the production process requires the cutting of circular disks from 1 m × 1 m sheets of steel. At present the disk-pressing machine is set to press out 16 disks of diameter 0.25 m from each sheet. You are asked whether it is possible to rearrange the cutting heads to save on wastage. There is also a need to cut disks of diameter 0.1 m from the same sheets. What would be the best arrangement of the cutting heads to minimise wastage? Is it possible to produce a mathematical formula for the maximum number of disks of radius r that can be cut from a sheet of given dimensions?

Problem statement

Given the size of disk and the dimensions of the sheet, *find* the most efficient cutting pattern and the maximum number of disks that can be cut from one sheet.

Formulate a mathematical model

We can list our factors as in Table 10.6.

Table 10.6

Description	Type	Symbol	Units
Length of sheet	Input parameter	l	m
Breadth of sheet	Input parameter	b	m
Radius of disk	Input parameter	r	m
Number of disks	Output variable	N	Integer
Wastage	Output variable	W	%

Assumptions

1 The disk cutters can cut cleanly and with perfect accuracy so that we can let the circles touch on our diagrams.
2 We shall examine patterns in which each circle (except those near an edge) touches

(a) four others ('four-point contact' or 'square arrangement') and
(b) six others ('six-point contact' or 'triangular arrangement').

Obtain the mathematical solution

The simplest pattern is the square arrangement with four-point contact as shown in Fig. 10.2. For the case mentioned above, we have $l = b = 1, r = 0.125$ and $N = 16$; so the wastage is

$$W = 1 - 16\pi(0.125)^2 \approx 21.5\%.$$

If the same pattern is used for the case $l = b = 1$, $r = 0.05$, then $N = 100$ and $W = 1 - 100\pi(0.05)^2 \approx 21.5\%$ as before. (Is this surprising?)

For general values of the parameters, we can form n columns of disks with this pattern if $b > 2nr$ (Fig. 10.3). If we increase b, another column becomes possible if b reaches $(2n + 2)r$. This means that n is the integer part of $b/2r$ which we write as $[b/2r]$, the square brackets denoting 'the integer part of'. An exactly similar argument applies to the rows, from which we deduce that the number of rows is $[l/2r]$; so the total number of disks is $N = [b/2r][l/2r]$ and the wastage is $W = (lb - [b/2r][l/2r]\pi r^2)/lb$.

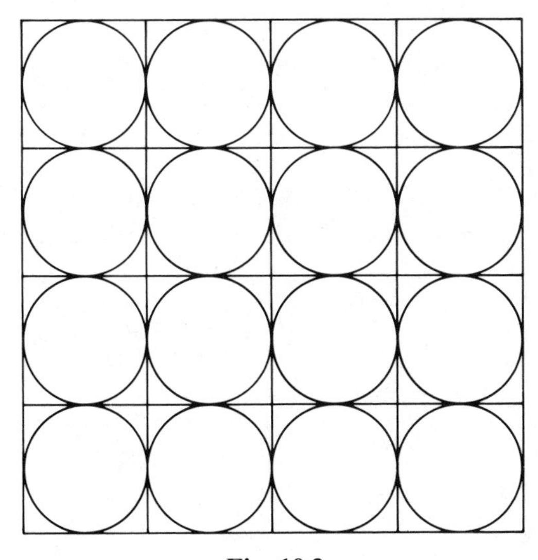

Fig. 10.2

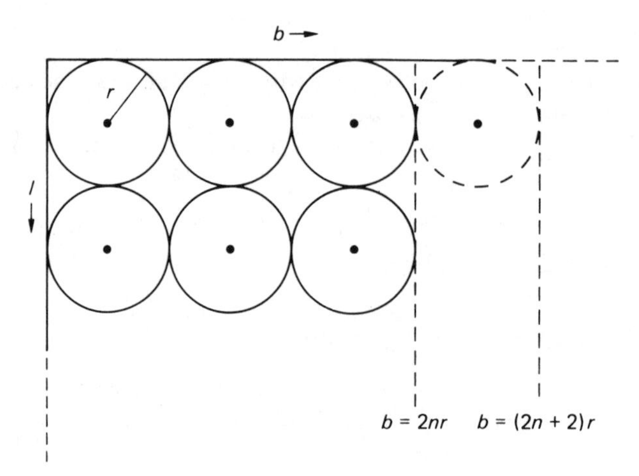

Fig. 10.3

Let us now look at the possibilities with six-point contact and general values of l, b and r. If we think of the pattern in terms of horizontal rows, it is clear from Fig. 10.4 that, when b is an odd-integer multiple of r (i.e. $(2n+1)r$), then each row can hold n disks and, if b is increased, this remains true until b reaches $(2n+2)r$ when an extra disk can be fitted in alternate rows. At this stage, we have alternate rows of $n+1$ and n disks. If b is increased further, the same pattern applies until we reach $b=(2n+3)r$ when equal rows of $n+1$ disks become possible.

Looking down the vertical side in Fig. 10.4, we see that, if there are x rows, then x must satisfy $2r+(x-1)r\sqrt{3}<l$ and, since x must be an integer, we can write $x=[1+(1/\sqrt{3})(l/r-2)]$. In the case of equal rows of n disks

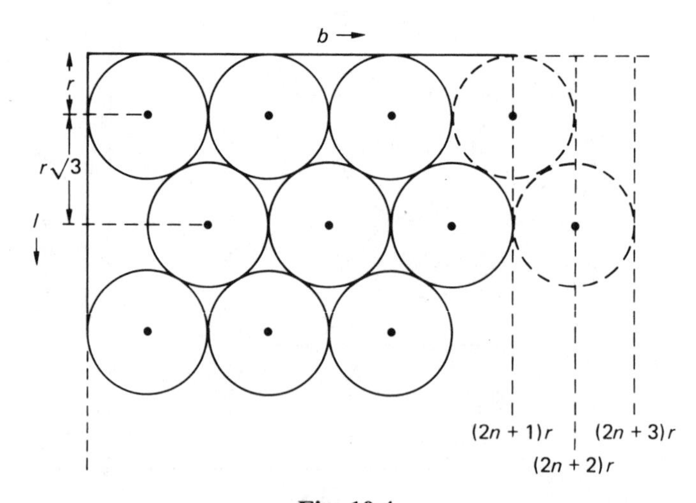

Fig. 10.4

the total number of disks is $N = nx$ and, since we have the condition $(2n+1)r \leqslant b < (2n+2)r$, we can write down the value of n as

$$n = \frac{1}{2}\left(\left[\frac{b}{r}\right] - 1\right).$$

So

$$N = \frac{1}{2}\left(\left[\frac{b}{r}\right] - 1\right)\left[1 + \frac{1}{\sqrt{3}}\left(\frac{l}{r} - 2\right)\right].$$

In the case of unequal rows of alternate $n+1$ and n disks the number of long rows (each with $n+1$ disks) is $x/2$ if x is even, and $(x+1)/2$ if x is odd. Also n has to satisfy $(2n+2)r \leqslant b < (2n+3)r$. Therefore,

$$n = \frac{1}{2}\left(\left[\frac{b}{r}\right] - 2\right).$$

The total number of disks is $N = x(n + \frac{1}{2})$ if x is even, and $N = x(n + \frac{1}{2}) + \frac{1}{2}$ if x is odd.

We can summarise the conclusions as follows. For equal rows of n disks when $(2n+1)r < b < (2n+2)r$,

$$N = \frac{x}{2}\left(\left[\frac{b}{r}\right] - 1\right).$$

For unequal rows of $n+1$ and n disks when $(2n+2)r < b < (2n+3)r$,

$$N = \begin{cases} \dfrac{x}{2}\left(\left[\dfrac{b}{r}\right] - 1\right), & x \text{ even,} \\ \dfrac{x}{2}\left(\left[\dfrac{b}{r}\right] - 1\right) + \dfrac{1}{2}, & x \text{ odd.} \end{cases}$$

In all cases,

$$x = \left[1 + \frac{1}{\sqrt{3}}\left(\frac{l}{r} - 2\right)\right].$$

We can think of N as a function of the two parameters l/r and b/r. A few sample values are given in Table 10.7. (Why do we start at $b/r = 3$ and $l/r = 4$?)

For comparison, the corresponding figures for the four-point contact (square) pattern are given in Table 10.8.

Interpret the mathematical solution

We see that there is no clear winner; either pattern can be the more efficient, depending on the values of the parameters. Note that, for parameter values in between the integer values tabulated, the value of N may remain the same but the wastage will alter.

Table 10.7

	N (six-point contact)					
l/r	$b/r=3$	$b/r=4$	$b/r=5$	$b/r=8$	$b/r=10$	$b/r=14$
4	2	3	4	7	9	13
7	3	5	6	11	14	20
10	5	8	10	18	23	33
15	8	12	16	28	36	52
20	11	17	22	39	50	72

Table 10.8

	N (four-point contact)					
l/r	$b/r=3$	$b/r=4$	$b/r=5$	$b/r=8$	$b/r=10$	$b/r=14$
4	2	4	4	8	10	14
7	3	6	6	12	15	21
10	5	10	10	20	25	35
15	7	14	14	28	35	49
20	10	20	20	40	50	70

For the case $l=b=1$, $r=0.05$, quoted above, the six-point contact method gives

$$x=\left[1+\frac{1}{\sqrt{3}}\left(\frac{1}{0.05}-2\right)\right]=[11.39\ldots]=11,$$

and $b/r=20$ (an *even* integer); so the case of unequal rows ($n=9$, $n+1=10$) applies and

$$N=\frac{11}{2}(20-1)+\frac{1}{2}=105.$$

The four-point contact pattern obviously gives 100 disks and is therefore inferior in this case.

Further thoughts

1 Can the figure of 105 be improved? If we use a mixed strategy of equal and unequal rows, we find that rows of 10, 9, 10, 9, 10, 9, 10, 9, 10, 10, 10 can be fitted, giving a total of 106 disks! A mixed strategy of this kind seems worth investigating.

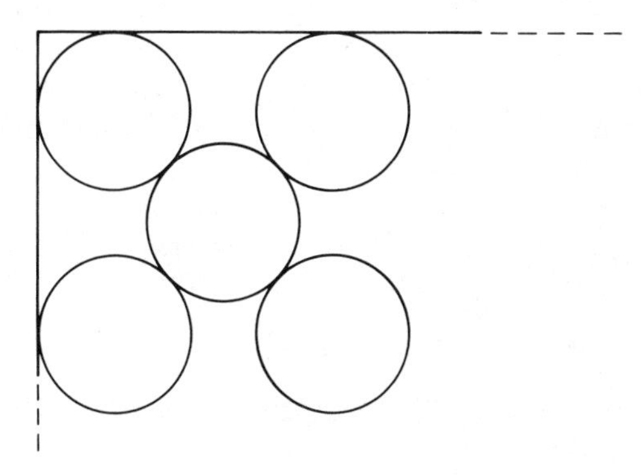

Fig. 10.5

2 Four-point contact does not have to be on a square arrangement. A staggered pattern can be adopted as shown in Fig. 10.5. Investigate the efficiency of such patterns.
3 Can you extend the model to deal with cases of two different sizes of disks which are cut from the same sheet. Under what conditions can the smaller circles be fitted in between the larger ones?

10.5 GUTTER

Context and problem statement

The building department of the local borough council require some specification on the size of gutters to fix to house roofs. In a particular new development, the roofs are all rectangular, 12 m long and 6 m from the ridge tile to the gutter. The angle of inclination of the roofs to the horizontal has not yet been decided, but will lie between 20° and 50°.

A guttering company is keen to get the contract from the council for the supply. The company say that their new durable plastic gutters will be sufficient and have always proved adequate no matter what the weather circumstances. The guttering cross-section is semicircular with radius 7.5 cm and the supplier claims that, for the roofs in question, one down drain-pipe of diameter 10 cm will be sufficient.

The council's chief housing expert is uncertain about the claims of the guttering supplier and so calls you in to set up a mathematical model so that a thorough analysis can be carried out before a bulk order for the plastic guttering is made. The housing chief is particularly interested in whether the size of the gutter will normally be sufficient to cope with a heavy fall of rain.

This investigation is concerned with the capacity of the gutter to hold rainwater. It is an input–output type of investigation that can occur in other situations such as the flow of water into and out of a water tank, river or reservoir. In this case the input comes from rainwater flowing in from the inclined roof and the output from the vertical drain-pipe(s). The critical question is whether the gutter can hold all the rainwater without overflow. This means that we are interested in the height of water in the gutter at a particular time. As the gutter cross-section is semicircular, then when the height of water is equal to the radius we shall get over-flow.

Formulate a mathematical model

Following the systematic plan for mathematical modelling as we have done in previous cases, the first job is to list the relevant factors (Table 10.9).

The problem can be clarified with the aid of the diagram in Fig. 10.6. We now apply a 'rate of flow' principle to the gutter system in the following form:

$$\left\{\begin{matrix}\text{volume rate of change}\\ \text{in water in the gutter}\end{matrix}\right\} = \left\{\begin{matrix}\text{inflow rate}\\ \text{from rain}\end{matrix}\right\} - \left\{\begin{matrix}\text{outflow rate down}\\ \text{the drain-pipe}\end{matrix}\right\}.$$

In terms of the notation given in Table 10.9, this is

$$V'(t) = Q_\text{i} - Q_\text{o}. \tag{10.6}$$

Note that Q_i and Q_o are *volume* flow rates while the rainfall rate r is a linear rate measured in metres per second.

Assumptions

We shall make a few minor assumptions concerning the flow.

1 All the rainwater falling on the roof will enter the gutter.
2 Rainwater falling directly into the gutter can be neglected.
3 There are no unforeseen blockages in the system due to leaves, etc.

Table 10.9

Description	Type	Symbol	Units
Rainfall rate	Input variable	r	m s^{-1}
Time	Variable	t	s
Angle of roof	Input parameter	α	deg
Length of roof	Input parameter	d	m
Width of roof (from ridge tile to gutter)	Input parameter	b	m
Radius of gutter	Input parameter	a	m
Height of water in the gutter	Output variable	h	m
Volume of water in the gutter	Variable	V	m^3
Inflow rate to gutter	Variable	Q_i	m^3 s^{-1}
Outflow rate from gutter	Variable	Q_o	m^3 s^{-1}
Cross-sectional area of the drain-pipe	Parameter	A	m^2
Acceleration due to gravity	Constant	g	m s^{-2}

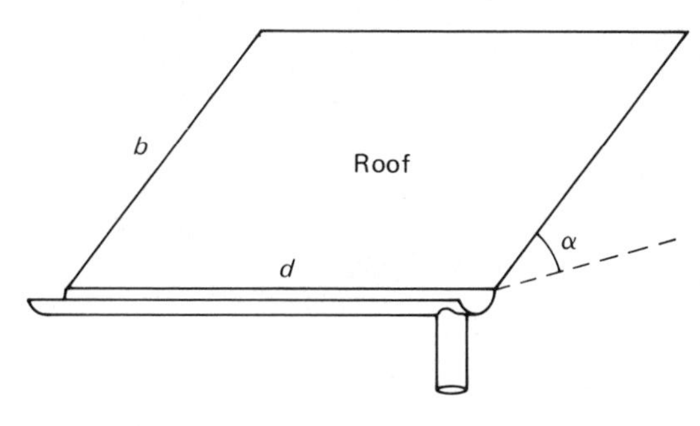

Fig. 10.6

4 Rain falls straight down onto the roof.
5 Rain does not splash away on hitting the roof.

The area of the roof is bd but, owing to the inclination, the area on which rain falls is $bd \cos \alpha$. As the rainfall rate is $r(t)$, the *volume* rate of rainfall on the roof is

$$r(t) \times \text{area} = r(t)bd \cos \alpha. \tag{10.7}$$

Also, the inclination of the roof will affect the flow rate into the gutter: the steeper the roof, the faster will be the flow of rain. Referring to Fig. 10.7, we see that the component of the rain's velocity down the roof is required and this introduces a $\sin \alpha$ term into our equation.

Hence an expression for the inflow of rainwater into the gutter is

$$Q_i = r(t)bd \sin \alpha \cos \alpha. \tag{10.8}$$

This is a simplification of what actually happens as rain is falling on top of water that is already flowing down into the gutter.

Now the volume of water lying in the gutter at any moment can be found by considering the cross-section of the gutter as shown in Fig. 10.8.

The volume is then found from

$$\text{volume} = \text{cross-section} \times \text{length } d \text{ of gutter.}$$

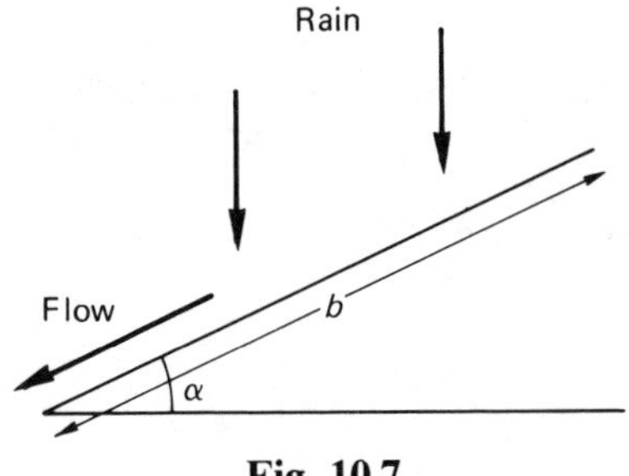

Fig. 10.7

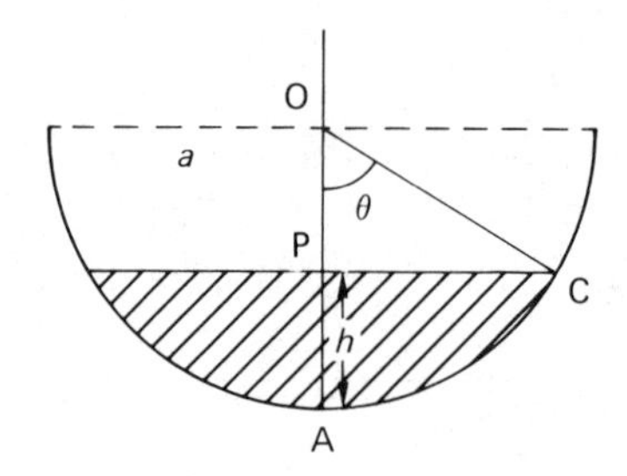

Fig. 10.8

The cross-section is a segment of a circle. From the diagram, using θ to denote the angle AOC, the segment area is given by

$$\begin{aligned}\text{area} &= \tfrac{1}{2}a^2 \times 2\theta - \tfrac{1}{2}a^2 \sin 2\theta \\ &= a^2(\theta - \sin\theta\cos\theta),\end{aligned}$$

but $\cos\theta = (a-h)/a$ and so $\sin\theta = \sqrt{2ah-h^2}/a$. Hence,

$$V(t) = da^2\left[\cos^{-1}\left(\frac{a-h}{a}\right) - \frac{(a-h)\sqrt{2ah-h^2}}{a^2}\right]. \tag{10.9}$$

What is in fact required is the rate of change of $V(t)$.

Now $V'(t) = (\mathrm{d}V/\mathrm{d}h)(\mathrm{d}h/\mathrm{d}t)$ and an expression for $\mathrm{d}V/\mathrm{d}h$ can be found from differentiating equation (10.9). After a little manipulation, we obtain

$$V'(t) = 2h'(t)d\sqrt{2ah-h^2}. \tag{10.10}$$

Now the outflow from the gutter is through the vertical drain-pipe. The velocity of the outflow is related to the height of water in the gutter, and the usual relation is obtained by applying Torricelli's law of energy conservation which states that potential energy lost over height $h(t)$ is balanced by the kinetic energy gained on exit down the pipe. Thus the exit velocity is $\sqrt{2gh}$ and, since the pipe cross-section is denoted by A, then the expression for the outflow is

$$Q_o = A\sqrt{2gh}. \tag{10.11}$$

If we return to equation (10.6), substituting for the flow rates gives the differential equation

$$r(t)bd\sin\alpha\cos\alpha - A\sqrt{2gh} = 2h'(t)d\sqrt{2ah-h^2},$$

i.e.

$$h'(t) = \frac{r(t)bd\sin\alpha\cos\alpha - A\sqrt{2gh}}{2d\sqrt{2ah-h^2}}. \tag{10.12}$$

Obtain the mathematical solution

A units check is useful for equation (10.12). Referring back to chapter 4, we can see that both sides have dimension L/T. Note also that an initial state is required (as is the case for any differential equation). Equation (10.12) is a first-order equation and a suitable initial condition might be $h(0)=0$, i.e. the gutter is dry and it starts to rain. There is a mathematical difficulty with this condition, however, in that equation (10.12) is 'singular' at $h=0$, i.e. we cannot evaluate $h'(0)$ when $h=0$ is substituted. This situation may be avoided by working with $\mathrm{d}t/\mathrm{d}h$ instead, or taking $h(0)=1.0$ cm, for example. What would be the effect of trying other starting values?

Some data have been given above for this model, but a full list is now convenient: $a=0.075\,\mathrm{m}$; $b=6.0\,\mathrm{m}$; $d=12.0\,\mathrm{m}$; $g=9.81\,\mathrm{m\,s^{-2}}$; $A=0.0025\pi\,\mathrm{m}^2$; $\alpha=30^\circ$. Substituting these values into equation (10.12) gives

$$h'(t)=\frac{1.299r-0.0145\sqrt{h}}{\sqrt{0.15h-h^2}}. \tag{10.13}$$

Solving the differential equation (10.13) will give us $h(t)$, the height of water in the gutter at any time. Our problem is now in the form, *given* the rainfall rate $r(t)$, *find* the depth $h(t)$.

Two possible expressions for the rainfall rate $r(t)$ are as follows:

$$r(t)=\text{constant},$$

or

$$r(t)=\begin{cases}\dfrac{1}{20}\sin\left(\dfrac{\pi t}{40}\right), & 0<t<40,\\ 0, & t>40.\end{cases}$$

The first model is the equivalent of steady persistent rain which sets in over a long period. Either the gutter overflows as it cannot cope, or a steady state will arise where the depth of water in the gutter $h(t)$ settles to a constant value of less than 0.075 m. From equation (10.13), the steady state can be mathematically predicted by examining what happens if $h'(t)=0$, which applies when there is a steady state:

$$1.299r=0.0145\sqrt{h}$$

$$h=8025.7r^2.$$

For example, $r=0.025\ \mathrm{cm\ s^{-1}}$ a steady-state situation occurs at $h\approx 5$ cm, as can be seen from Table 10.10 and from the graph (Fig. 10.9).

For the second model (illustrated in Fig. 10.10) the rain profile represents a short heavy burst which rises to a peak at 20 s and then decreases to zero at 40 s. Thus we would expect the gutter to fill up rapidly before the level falls off. This is the behaviour predicted by the model when equation (10.13)

Table 10.10

t/s	0	5	10	15	20	25	30	35
h/cm	1.00	2.39	3.11	3.58	3.92	4.17	4.35	4.50
t/s	40	45	50	55	60	...	120	
h/cm	4.61	4.69	4.76	4.81	4.86	...	5.00	

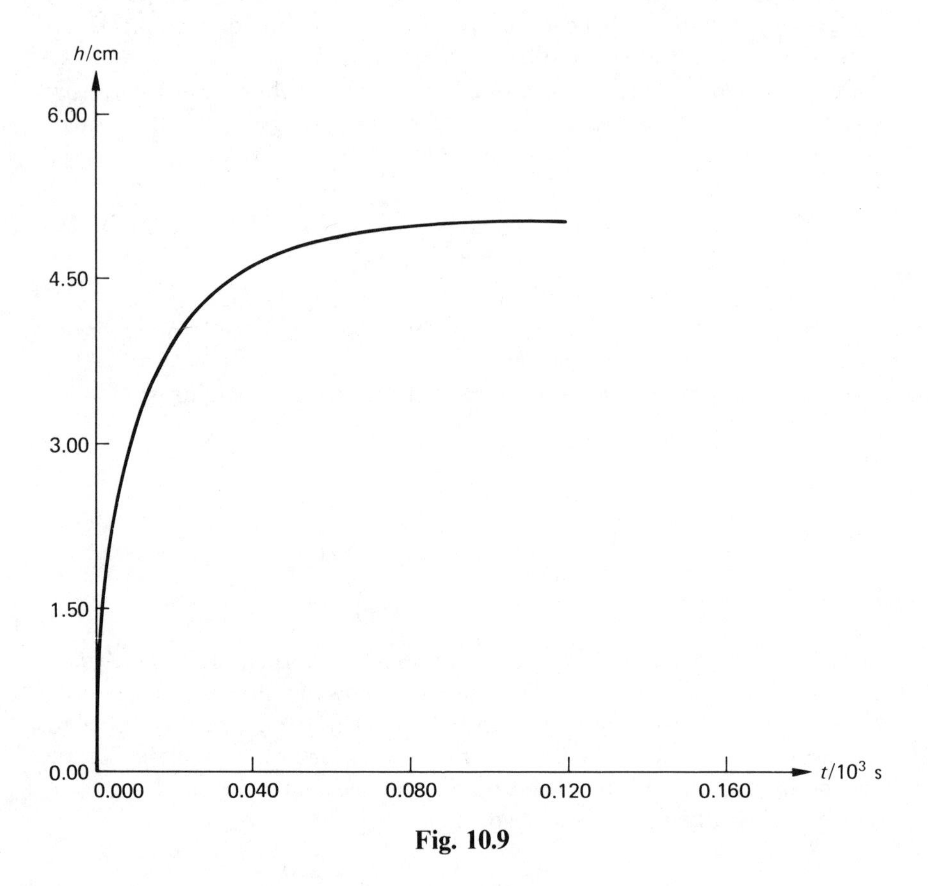

Fig. 10.9

is rewritten as

$$h'(t) = \begin{cases} \dfrac{1.299 \times \frac{1}{20}\sin(\pi t/40) - 0.0145\sqrt{h}}{\sqrt{0.15h - h^2}}, & 0 < t < 40, \\[2ex] \dfrac{-0.0145\sqrt{h}}{\sqrt{0.15h - h^2}}, & 40 < t. \end{cases}$$

The results of solving for $h(t)$ are shown in Table 10.11.

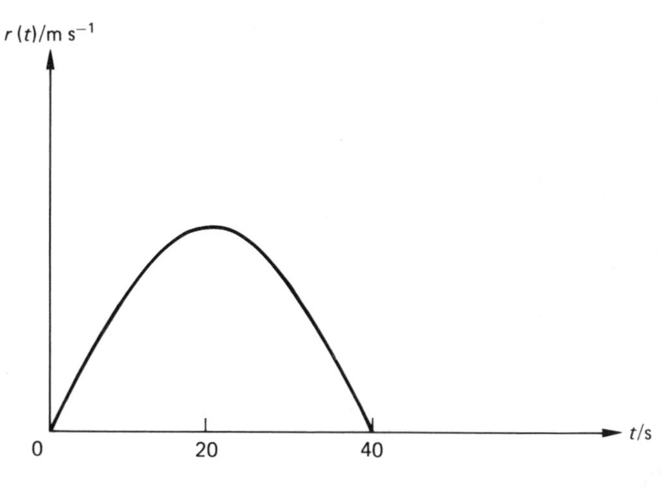

Fig. 10.10

Table 10.11

t/s	0	2	4	6	8	10	12	14	
h/cm	1.0	0.28	0.12	0.39	0.84	1.45	2.19	3.03	
t/s	16	20	25	30	35	40	45	50	51
h/cm	3.92	5.70	7.35	7.65	6.42	4.30	2.74	0.23	Empty

Figure 10.11 on p. 234 shows the graph obtained for h against t and, as we would expect, the response of the gutter follows the rain profile, rising to a maximum at around $t = 28$ s and then subsiding to zero after about 51 s. Again a slight mathematical problem occurs near the end because, as h becomes very small, we are almost dividing by zero in our differential equation for $h(t)$.

10.6 TURF

Context

Grass court tennis matches are always susceptible to interruption due to rain. A water-proof cover is not often available and so play can only be resumed when the top layer of turf has dried out sufficiently. This means that either the rainwater has soaked right through to the subsoil, or it has evaporated back into the atmosphere after it has stopped raining. A number of mechanical devices can be tried to speed up the drying process but, to avoid damaging the turf, it is often best to let the turf dry out naturally. Can a mathematical model be set up to represent the drying process?

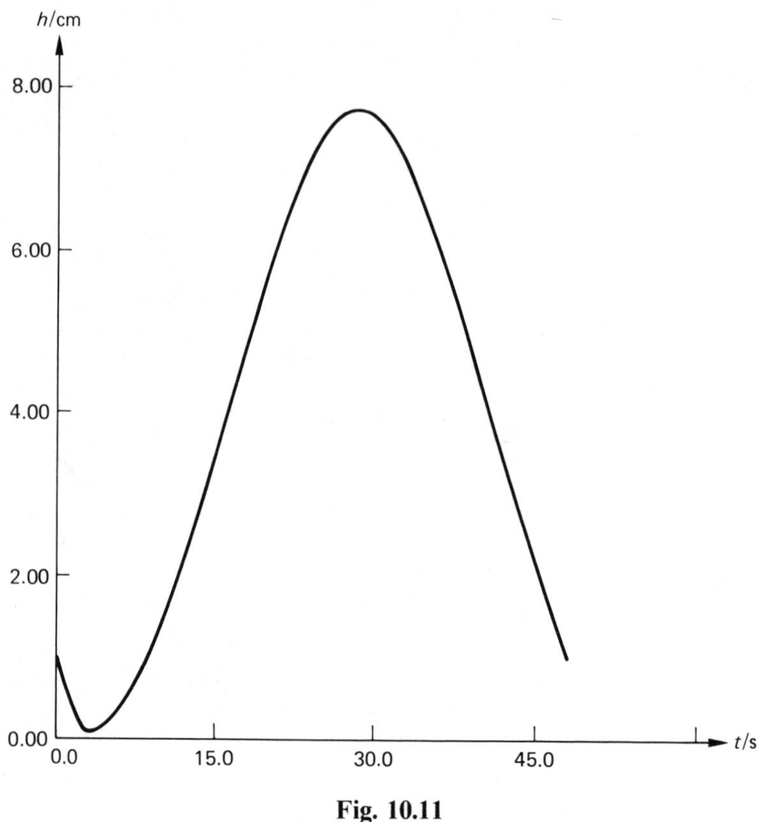

Fig. 10.11

Problem statement

Given a local rain shower, is it possible to predict when play can be resumed? In particular, suppose that the turf is dry to begin with when it suddenly starts to rain and continues at a constant rate of half an hour. Suppose also that the amount of rain collected in this time is 1.8 cm. (Note that this is a depth and not a *volume* until you multiply by the area of catchment.)

Formulate a mathematical model

The problem is somewhat like that in section 10.5 in that we have a flow principle again:

$$\left\{\begin{matrix}\text{rate of increase in}\\ \text{water in the turf}\end{matrix}\right\}=\left\{\begin{matrix}\text{inflow}\\ \text{rate}\end{matrix}\right\}-\left\{\begin{matrix}\text{outflow}\\ \text{rate}\end{matrix}\right\}.$$

We list the factors in Table 10.12.

Note that Q is measured in metres and becomes an actual volume of water when multiplied by the area of the turf. Similarly e and s are measured in

Table 10.12

Description	Type	Symbol	Units
Rainfall rate	Variable	$r(t)$	$m\ s^{-1}$
Time	Variable	t	s
Area of turf	Parameter	A	m^2
Thickness of turf	Parameter	D	m
Quantity of rain currently in the turf	Variable	$Q(t)$	m
Evaporation rate	Variable	$e(t)$	$m\ s^{-1}$
Soak-away	Variable	$s(t)$	$m\ s^{-1}$
Proportionality constants		a, b	s^{-1}
Time at which it stops raining	Parameter	c	s

metres per second and, when multiplied by A, then become *volume* flow rates measured in cubic metres per second.

Now consider the model in Fig. 10.12. Since the turf is dry to begin with, we have an initial condition $Q(t=0)=0$. With reference to the 'flow principle' above, it is necessary to model the inflow, outflow and turf holding capacity. The inflow is easily done as it is the product of the rainfall rate and the area of turf considered. Hence

$$\text{inflow rate} = r(t)A. \qquad (10.14)$$

For the outflow rate, we have to decide how the water will disappear from the turf. This will be through drainage into the subsoil, and through evaporation back into the atmosphere. While it is still raining, it is unlikely that there will be any evaporation; so only the soak-away rate has to be modelled. We shall take it to be proportional to the amount of water currently present in the turf, i.e.

$$s(t) = aQ(t), \qquad (10.15)$$

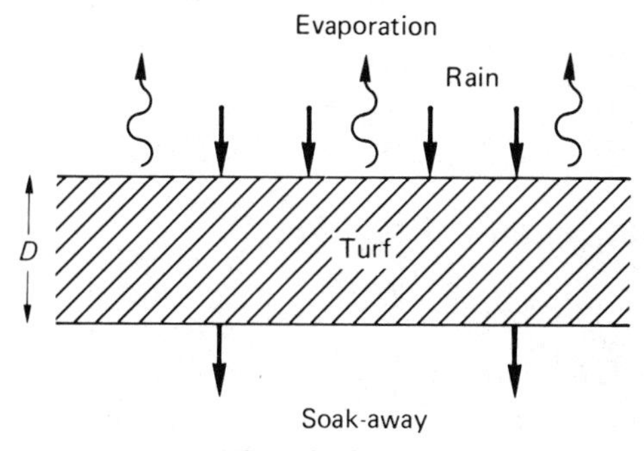

Fig. 10.12

Once it has stopped raining, water continues to soak away through the subsoil and can now also evaporate. The rate of evaporation will depend on the local air temperature and humidity. Also the water that evaporates will come from the top surface of the wet turf. To keep the model simple, the evaporation rate will be taken as proportional to the amount of water in the turf, i.e.

$$e(t) = bQ(t).$$

Now a differential equation can be formed for $Q(t)$ and is expressed in two stages according to whether it is still raining or not:

$$\dot{Q}(t) = \begin{cases} r(t) - aQ(t), & 0 < t < c, \\ \quad - aQ(t) - bQ(t), & c < t. \end{cases} \tag{10.16}$$

This model could be made more realistic but also more complicated by considering alternative expressions for the soak-away and evaporation rates.

To see whether the solution of equation (10.16) gives an adequate result for the water held in the turf, we obtain $Q(t)$ by integrating the equation, first supplying data for $r(t)$, a and b. Note in passing that the turf thickness D has not come into the problem. The rainfall rate $r(t)$ was stated earlier to be constant over a period of half an hour, i.e.

$$c = 1800 \text{ and } r(t) = \frac{0.018}{1800} = 10^{-5} \text{ m s}^{-1}.$$

To proceed with the development of the model, we must put in suitable values for the proportionality constants a and b, but what values are 'suitable'? At this point, which very often occurs in modelling, we have to choose from a number of alternative courses of action. One is to look deeper into the theoretical principles involved and to hope to find appropriate values of a and b from scientific considerations. Another option is to continue with a and b as parameters with unspecified values, to obtain the mathematical solution and then to compare the model's predictions with real data. It will then be a question of deriving the values of a and b for the best fit to the data (see chapter 6). For the purpose of obtaining a specific answer for this example, we shall avoid these issues here and insert particular values of $a = 0.001 \text{ s}^{-1}$ and $b = 0.0005 \text{ s}^{-1}$.

Obtain the mathematical solution

From equation (10.16), if we take 1 m^2 of turf then $A = 1$ and the differential equations to be solved are

$$\dot{Q}(t) = \begin{cases} 10^{-5} - 10^{-3}Q(t), & 0 < t < 1800, \quad (10.17) \\ \quad - 10^{-3}Q(t) - 5 \times 10^{-4}Q(t), & t > 1800. \quad (10.18) \end{cases}$$

Both these are simple linear first-order equations and the solution can be obtained without difficulty. Integrating equation (10.17) gives

$$Q(t) = 0.01[1 - \exp(-0.001t)]. \tag{10.19}$$

This holds up to $t = 1800$, at which time $Q(t) \approx 0.008\,35$.

Now solving equation (10.18) in the same way gives

$$Q(t) = B\exp(-0.0015t),$$

where B is an integration constant to be calculated from the condition $Q(1800) = 0.008\,35$. Substituting, we get $B = 0.124$. Hence,

$$Q(t) = 0.124\exp(-0.0015t), \qquad t > 1800. \tag{10.20}$$

Interpret the solution

Equation (10.20) predicts how the quantity of water in the turf decreases after the end of the shower. Our problem was to find when play can be resumed, which presumably means when the turf is dry again; so $Q = 0$. However, the expression in equation (10.20) is a negative exponential function and $Q(t)$ will never actually be zero according to this model. We can make an assumption that, when the water quantity has dropped to 10% of its peak value, then the turf is dry enough. How long we have to wait is given by putting $Q = 0.000\,835$ in equation (10.20), i.e. $0.000\,835 = 0.124\exp(-0.0015t)$. Taking logarithms, we find that $-7.088\,08 = -2.0875 - 0.0015t$, giving $t = 3334$. So, after the end of the shower, we shall have to wait a further 1534 s or about 16 min.

Our choice of 10% was rather arbitrary. How much longer would it take to get down to 5% of the maximum? With $Q = 0.004\,175$ in equation (10.20) we find that $t = 3796$; so a further 462 s or about 8 min is required.

10.7 PARACHUTE JUMP

Context and problem statement

When a parachute jump is made from an aeroplane, quite often the first part of the descent is made in 'free fall' at a high altitude. After a certain time the parachutist then opens the parachute to enable a soft landing to be completed. The important question for the parachutist is to decide when the parachute should be opened. This must obviously not be too late; otherwise the landing speed will be too great, resulting in injury or death of the performer. On the other hand, at a great altitude with a thin atmosphere a free fall can generate high speeds, which can add to the exhilaration of the event. There may also be reasons for not opening the parachute immediately owing to the proximity of the aeroplane and other parachutists. It is therefore interesting to model the motion of the parachute jump and to see whether there is some optimum position at which the parachute should be opened.

Suppose that the aeroplane is travelling horizontally at a height of 500 m with a speed of 125 m s^{-1}. The mass of the person jumping together with the parachute seems like a data requirement, but it turns out that it is more important for the simulation to make accurate assessments of the air resistance and the obvious change in air drag when the parachute is opened. The idea of terminal speed is important again (see sections 4.4 and 8.3). The terminal speed of a human in free fall will be taken as 120 miles h^{-1}, as before. Now obviously the parachutist needs to make a soft landing; from information concerning jumps the usual comment is that a parachute landing should be like falling off a 12 ft wall. This means that the terminal velocity with the parachute open must be the same as the velocity acquired on falling off the wall. Now, from simple energy considerations on falling through a height h (and neglecting air resistance over a small distance), the velocity v is given by $v^2 = 2gh$. Here $h = 12$ ft $= 12 \times 12/39.37$ m. Also $g = 9.8065$ m s^{-2}; so $v^2 = 2 \times 9.8065 \times 12 \times 12/39.37$, i.e. $v \approx 8.47$ m s^{-1}.

If we allow for air resistance proportional to the square of velocity, then, following the notation in section 8.3, the air resistance constant k is given by

$$\begin{aligned} k &= \frac{g}{v^2} \\ &= \frac{9.8065}{(8.47)^2} \\ &\approx 0.1367\ (\text{m}^{-1}). \end{aligned} \tag{10.21}$$

The air resistance constant for the free-fall part of the drop can be calculated from the terminal velocity formula $v = g/k$, which is the case when the air resistance is taken directly proportional to velocity (again see section 8.3). This is valid since the atmosphere is fairly thin at a great height and the falling human is relatively compact in shape. If we take the terminal velocity to be 120 miles h^{-1} as stated above, this converts to 53.645 m s^{-1}. Hence,

$$\begin{aligned} k &= \frac{9.8065}{53.645} \\ &\approx 0.1828\ (\text{s}^{-1}). \end{aligned} \tag{10.22}$$

The equations of motion for the parachutist will be very similar to those in section 8.3 and the same notation can be followed.

Formulate a mathematical model

The diagram in Fig. 10.13 helps to clarify the problem and define the required variables. The list of factors is given in Table 10.13.

Application of Newton's law of motion gives two equations: one for the

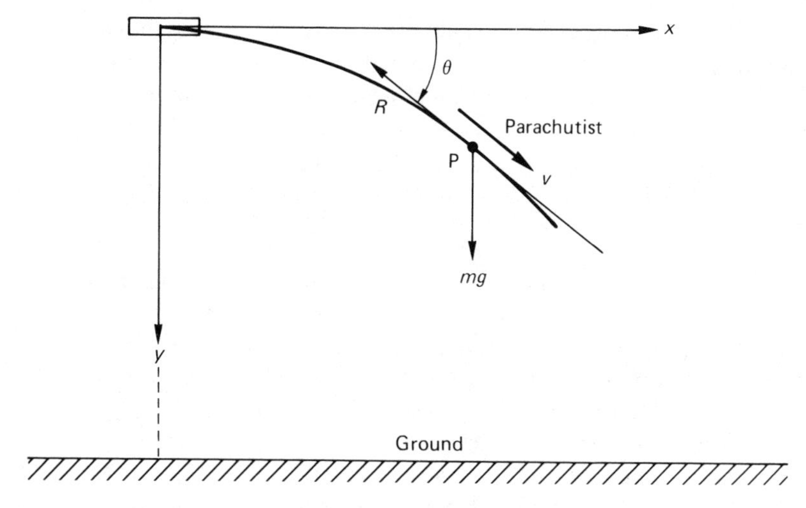

Fig. 10.13

Table 10.13

Description	Type	Symbol	Units
Horizontal distance travelled by parachutist	Variable	$x(t)$	m
Vertical distance dropped by parachutist	Variable	$y(t)$	m
Time	Variable	t	s
Velocity of parachutist	Variable	v	m s^{-1}
Mass of parachutist plus equipment	Parameter	m	kg
Air resistance force	Variable	R	N
Air resistance constant	Parameter	k	s^{-1}*
Acceleration due to gravity	Constant	g	m s^{-2}
Angle of inclination of path of parachutist	Variable	θ	deg
Initial height (given)	Parameter	h	m
Initial horizontal velocity (given)	Parameter	u	m s^{-1}

* Provided that R is directly proportional to v.

horizontal motion and one for the vertical motion. For horizontal motion, we have

$$-R\cos\theta = m\frac{d^2x}{dt^2} = m\ddot{x} \tag{10.23}$$

and, for vertical motion,

$$-R\sin\theta + mg = m\frac{d^2y}{dt^2} = m\ddot{y}. \tag{10.24}$$

Now $R = mkv^n$, $v \cos\theta = \dot{x}$ and $v \sin\theta = \dot{y}$, where $\dot{x}$ and $\dot{y}$ are the components of velocity, and the exponent n will be taken as 1 for free fall and 2 for the effect once the parachute has opened. Also from the components, we have the relation $v^2 = \dot{x}^2 + \dot{y}^2$. We can therefore eliminate v and θ to give

$$-k\dot{x}(\dot{x}^2 + \dot{y}^2)^{(n-1)/2} = \ddot{x}, \tag{10.25}$$

$$-g - k\dot{y}(\dot{x}^2 + \dot{y}^2)^{(n-1)/2} = \ddot{y}. \tag{10.26}$$

These two coupled simultaneous differential equations make up the general mathematical model for the parachute drop. They cannot be solved without a computer program to obtain approximate answers. In passing, note that the main assumption made so far is that the motion takes place entirely *in one plane*, denoted here by the axes Oxy. Now, equations (10.25) and (10.26) are non-linear since they contain powers of $\dot{x}$ and $\dot{y}$. However, for the first part of the motion, as we have already stated, if the assumption is made that the air resistance is directly proportional to velocity, then the exponent $n = 1$. This leads to some simplification of the equations so that, in fact, equations (10.25) and (10.26) can be integrated exactly.

Obtain the mathematical solution

With $n = 1$ for the first part of the motion, the equations to be solved are

$$\ddot{x} = -k\dot{x}$$

and

$$\ddot{y} = -k\dot{y} + g,$$

where, from the data above, the value of k will be taken as 0.1828. Initial conditions are as follows: $x(0) = 0$ and $\dot{x}(0) = u = 125$; $y(0) = 0$ and $\dot{y}(0) = 0$. Integrating twice in each case, it is easy to obtain the solutions

$$x(t) = \frac{u}{k}[1 - \exp(-kt)] \tag{10.27}$$

and

$$y(t) = \frac{gt}{k} - \frac{g}{k^2}[1 - \exp(-kt)]. \tag{10.28}$$

So long as the assumptions on air resistance hold and the parachute is not opened, then these two equations give the path of the falling person. In fact, y can be explicitly expressed in terms of x by eliminating t between equations (10.27) and (10.28) to give

$$y = -\frac{gx}{ku} - \frac{g}{k^2}\ln\left(1 - \frac{kx}{u}\right). \tag{10.29}$$

A units check (see chapter 4) may be carried out so that, for example, we can be sure that a k has not been dropped in error. Also the well-known results for projectile motion *in vacuo* can be obtained from equation (10.29) by expanding it in a series and then putting $k=0$. This is left for our readers to try out.

It is interesting to chart the velocity of the parachutist as the drop takes place. This can be done from the velocity equations obtained by differentiating equations (10.27) and (10.28):

$$\dot{x}(t) = u \exp(-kt),$$

$$\dot{u}(t) = \frac{g}{k}[1 - \exp(-kt)].$$

Therefore,

$$v^2 = u^2 \exp(-2kt) + \left(\frac{g}{k}\right)^2 [1 - 2\exp(-kt) + \exp(-2kt)]. \quad (10.30)$$

If a graph is drawn of the behaviour of v as time elapses, it can be seen that v has a minimum value. This surprising result can be checked by differentiation:

$$\frac{\mathrm{d}(v^2)}{\mathrm{d}t} = -2ku^2 \exp(-2kt) + \left(\frac{g}{k}\right)^2 [2k \exp(-kt) - 2k \exp(-2kt)].$$

$\mathrm{d}(v^2)/\mathrm{d}t = 0$, for a minimum value of v^2 and also for v. Simplifying, we get

$$t = \frac{1}{k} \ln\left(\frac{g^2 + k^2u^2}{g^2}\right). \quad (10.31)$$

The speed at this time is easily obtained from equation (10.30) as

$$v^2 = \left(\frac{g^2 u}{g^2 + k^2u^2}\right)^2 + \left(\frac{g}{k}\right)^2 \left(1 - \frac{g^2}{g^2 + k^2u^2}\right),$$

which simplifies to

$$v = \frac{gu}{\sqrt{g^2 + k^2u^2}}. \quad (10.32)$$

Now the initial value of v is u and the terminal speed is g/k; so equation (10.32) can be checked against these two values to show that the speed of the parachutist reaches a minimum value, less than both the initial and the terminal velocities. On examining the inequalities, we can easily show that

$$\frac{gu}{\sqrt{g^2 + k^2u^2}} < u \text{ (initial speed)}$$

and

$$\frac{gu}{\sqrt{g^2 + k^2u^2}} < \frac{g}{k} \text{ (terminal speed).}$$

From the data given, calculations using equations (10.31), (10.32), (10.27) and (10.28) can be carried out to show that in this case the parachutist reaches a minimum speed of 49.31 m s^{-1} after 10.18 s, when his position is $x = 577.5$ m and $y = 298.4$ m.

If we return to equations (10.25) and (10.26), it is possible to get computer solutions for these equations for general values of k and n. For instance, if the 'linear' model for free fall is rejected, and instead we take $n = 2$, then an alternative value of k is easily calculated from the terminal velocity relation $g = kv^2$. It turns out that the parachutist *still experiences decreasing velocity* to some minimum value before speeding up again.

When the parachute is opened, there will be a sharp jerk due to the abrupt change in the resistance force. This force can be considerable (and probably unpleasant for the parachutist), which is why the decision can now be made to open the parachute when the speed is a minimum. We can calculate this change in force which causes the jerk from the difference between the values of R before and after the opening. Before the opening,

$$R = mk_1v = m \times 0.1828 \times 49.31$$

and, after the opening,

$$R = mk_2v^2 = m \times 0.1367 \times (49.31)^2.$$

The difference is about $300m$ (N) which is large, but on the other hand the

Table 10.14

t/s	y/m	Speed/m s^{-1}	Comment
0	0	125.00	Starting conditions
2	17.43	91.96	
4	62.37	66.29	Free fall
6	126.41	54.95	
8	203.69	50.37	Drag $\propto$ speed
10	290.16	45.02	Parachute opened
11	323.49	19.54	Parachute effect
12	333.33	8.79	$\dot{x}$ negligible
13	341.89	8.50	Path is almost vertical
14	350.37	8.47	Drag $\propto$ (speed)^2
15	358.84	8.47	Terminal speed
⋮	⋮	⋮	
33	500.00	8.47	Landing

parachute will not be opened instantaneously; so this change may be spread over several seconds.

To simulate the remainder of the descent, equations (10.25) and (10.26) must be solved in the case where $k = 0.1367$ and $n = 2$. As the height of the jump was taken to be 500 m, there is about 200 m left to fall. Data from the computer simulation are given in Table 10.14.

Figure 10.14 shows the path of the parachutist and Fig. 10.15 shows how the speed varies with time.

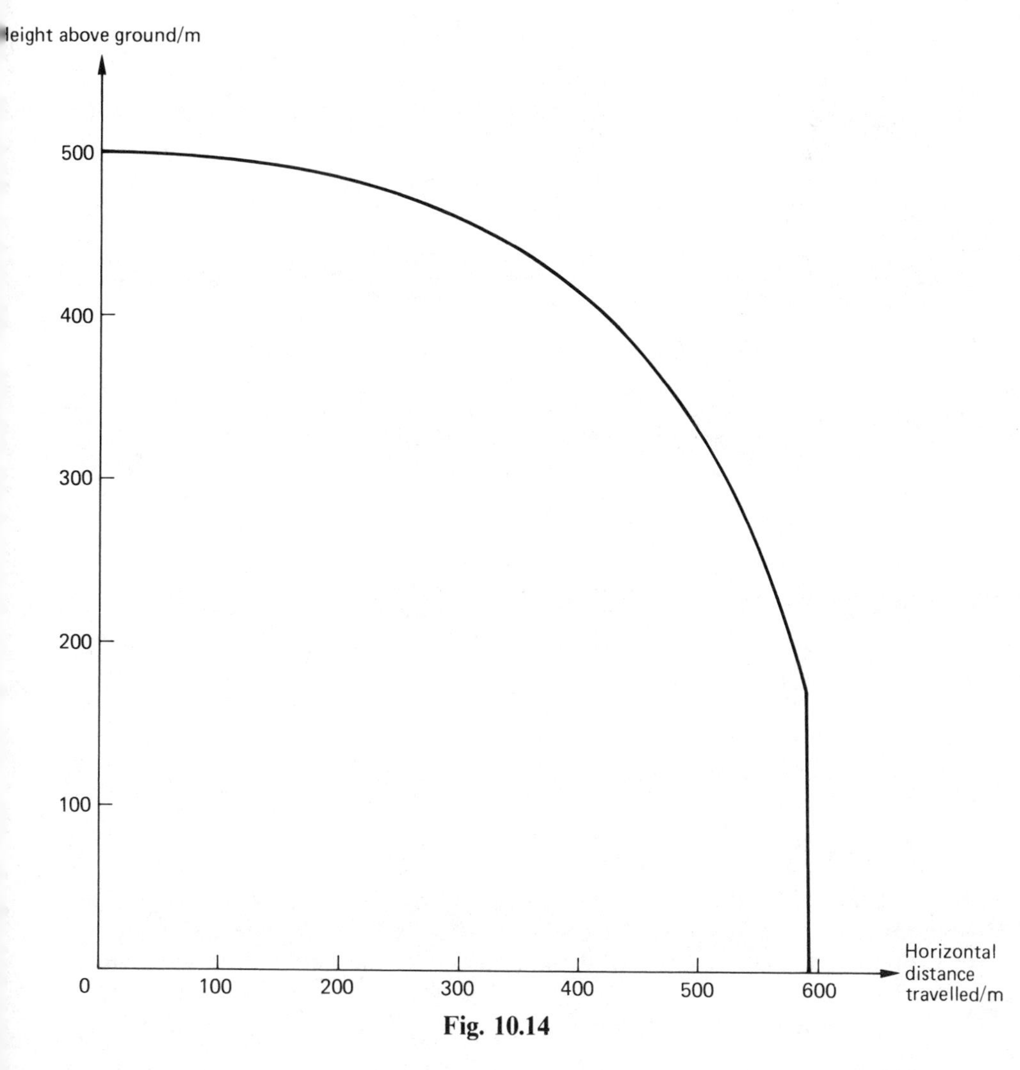

Fig. 10.14

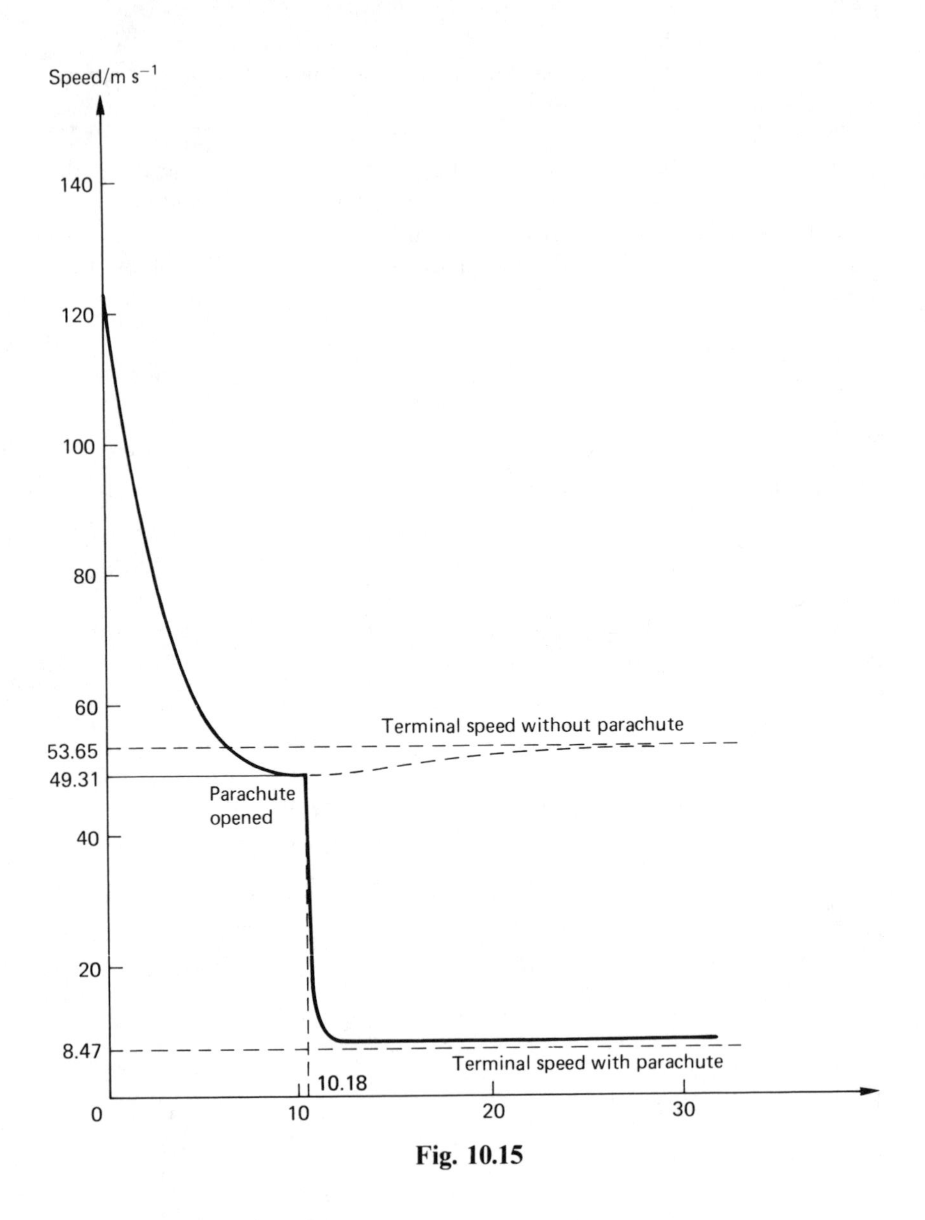

Fig. 10.15

10.8 ON THE BUSES

Context

Bus timetables are very difficult to put into reliable operation because of the many random and irregular factors which affect the running of the buses. Suppose that, as part of a study into the operation of driver-operated buses, you are asked to develop a model to look into the problem of 'bunching' of buses on the same regular route.

Problem statement

What should be the time interval between buses leaving the bus depot so that

(a) buses on the route do not catch up with and/or overtake each other and
(b) passengers do not have very long waits in between buses?

Formulate a mathematical model

This problem is clearly a candidate for the stochastic simulation treatment described in chapter 7. We can develop, however, a simple *deterministic* model as a starting point, and the random elements can be introduced at a later stage. There are obviously three groups of factors involved: the buses, the bus stops and the passengers. So we can construct our factor list as follows.

Factors concerning the buses.
Time of leaving the depot.
Time of arrival at any stop.
Number of passengers leaving at any stop.
Time spent at a stop.
Total number of passengers carried.
Maximum capacity.
Speed of travelling.
Traffic conditions.

Factors concerning the bus stops.
Position of stop on the bus route.
Time elapsed since the last bus left.
Rate of arrival of passengers.
Number of passengers waiting when next bus arrives.
Distance between stops.

Factors concerning the passengers.
Time of arrival at a stop.
Length of journey (number of stops).
Waiting time for a bus.
Total journey time.

There is obviously some overlap here and we need to construct a convenient notation so that we can express the connections between the factors and reduce the number of variables.

For simplicity, consider a circular route with stops at regular 0.25 mile intervals (Fig. 10.16). If we ignore traffic delays and assume that the buses travel at a constant 30 miles h^{-1}, then all the journey times between stops will be 30 s. Assume that passengers arrive at any stop at an average rate of one per minute. The driver has to collect fares as passengers board his bus;

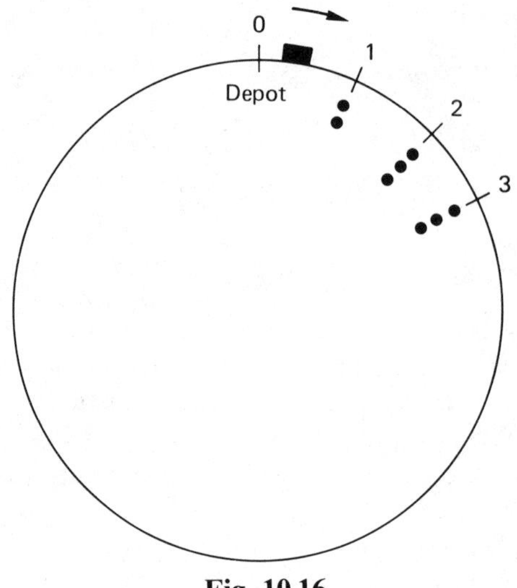

Fig. 10.16

assume that it takes a constant 10 s per passenger. Also assume that passengers take 5 s each to get off the bus and that the bus can carry up to a maximum of 60 passengers.

Suppose that a total number N_b of buses leave the depot at regular intervals T with the first leaving at time 0 and the Ith bus leaving at time $(I-1)T$. Let $T_a(I, J)$ be the time of arrival of bus I at stop J, $T_d(I, J)$ the time of departure of bus I from stop J, $N_1(I, J)$ the number of passengers leaving bus I at stop J and $N_e(I, J)$ the number of passengers entering bus I at stop J.

The time spent at stop J by bus I (which equals $T_d(I, J) - T_a(I, J)$) is clearly determined by the maximum of $\{5N_1(I, J), 10N_e(I, J)\}$. If we assume that the number of passengers waiting at stop J when bus I arrives is proportional to the time elapsed since the previous bus $I-1$ left, we have

$$N_e(I, J) = INT\left(\frac{T_a(I, J) - T_d(I-1, J)}{60}\right)$$

i.e.

$$N_e(I, J) = INT\left(\frac{T_d(I, J-1) + 30 - T_d(I-1, J)}{60}\right)$$

for an arrival rate of one per minute where $INT(x)$ stands for the integer part of x and is an alternative to the [ ] notation used in section 10.3. The first bus is rather special in this regard; so we can define $T_d(0, J) = 0$ for all J.

The number of passengers leaving bus I at stop J has to be worked out. We may have some data on the lengths of journeys made by passengers (Table 10.15).

In reality, there could well be a different distribution for each stop and for different times of the day. For our deterministic model, we can take a constant

Table 10.15

Number of stops	Percentage of passengers
1	8
2	12
3	20
4	30
5	20
>5	10

mean value for all passengers, say three stops. So, when $N_e(I,J)$ passengers board bus I at stop J, this gives rise to $N_l(I,J+3)=N_e(I,J)$. The total number of passengers on board the bus at any time will be given by the sum $\sum_J N_l(I,J)$ and we need to check that this does not exceed 60. We now have the beginnings of a simple model. We see that we need two arrays, namely $T_d(I,J)$ and $N_l(I,J)$, and clearly

$$T_d(I,J)=T_d(I,J-1)+30+\max\{10N_e(I,J),5N_l(I,J)\}.$$

For our convenience, we have assumed that no passengers arrive while a bus is actually at a stop. We have also assumed that we do not have buses overtaking each other. Since a study of overtaking was one of our objectives, we must build this feature into the model. As we have assumed a constant travelling time, any overtaking that occurs must happen at a stop. Once overtaking has occurred, the bus numbers become mixed up. We need an array $P(K)$ such that $P(K)$ is the original number of the bus which is currently Kth in the sequence. Initially, we have $P(K)=K$ for $K=1$ to N_b. The Kth bus in the sequence will arrive at stop J before the $(K-1)$th bus has left if

$$T_d(P(K),J-1))+30<T_d(P(K-1),J).$$

We call this 'bunching' and we can use a variable B to record the number of occurrences. When it happens, we can assume for convenience that all waiting passengers continue to board the bus in front so that $N_e(P(K),J)=0$. The condition for overtaking is that $T_d(P(K),J)<T_d(P(K-1),J)$. (We must define $P(0)=0$.) When it happens, we must interchange the number labels of the Kth and $(K-1)$th buses, i.e. $R=P(K)$, $P(K)=P(K-1)$ and $P(K-1)=R$. Note that we assume no 'double overtaking'.

In spite of all our simplifications, our model is already getting rather complicated and we have not yet considered the passengers! We need variables such as $T_{ar}(I,J)$ which is the time of arrival of the Ith passenger currently at stop J, $W(I,J)$ which is the waiting time of the Ith passenger at stop J, and $T_j(I,J)$ which is the total journey time of the Ith passenger at stop J and equals $W(I, J)$ + boarding time + travel time + disembarking time.

For a more realistic stochastic model, we can use a Poisson process to generate the arrivals of passengers at a stop and to sample the journey lengths from a distribution such as that mentioned earlier. We can also introduce random delays into the journey times between stops.

Obtain the mathematical solution

A run of the simple deterministic model on a microcomputer using only four stops and four buses with intervals of 500 s between starting times from the depot and 60 s between passenger arrivals at any stop gave the following results:

```
AT STOP 1
BUS 1 ARRIVED AT 30    LEFT AT 30    TOOK ON 0  DUMPED 0
BUS 2 ARRIVED AT 530   LEFT AT 610   TOOK ON 8  DUMPED 0
BUS 3 ARRIVED AT 1030  LEFT AT 1100  TOOK ON 7  DUMPED 0
BUS 4 ARRIVED AT 1530  LEFT AT 1600  TOOK ON 7  DUMPED 0

AT STOP 2
BUS 1 ARRIVED AT 60    LEFT AT 70    TOOK ON 1  DUMPED 0
BUS 2 ARRIVED AT 640   LEFT AT 730   TOOK ON 9  DUMPED 8
BUS 3 ARRIVED AT 1130  LEFT AT 1190  TOOK ON 6  DUMPED 7
BUS 4 ARRIVED AT 1630  LEFT AT 1700  TOOK ON 7  DUMPED 7

AT STOP 3
BUS 1 ARRIVED AT 100   LEFT AT 110   TOOK ON 1  DUMPED 1
BUS 2 ARRIVED AT 760   LEFT AT 860   TOOK ON 10 DUMPED 9
BUS 3 ARRIVED AT 1220  LEFT AT 1280  TOOK ON 6  DUMPED 6
BUS 4 ARRIVED AT 1730  LEFT AT 1800  TOOK ON 7  DUMPED 7

AT STOP 4
BUS 1 ARRIVED AT 140   LEFT AT 160   TOOK ON 2  DUMPED 1
BUS 2 ARRIVED AT 890   LEFT AT 1010  TOOK ON 12 DUMPED 10
BUS 3 ARRIVED AT 1310  LEFT AT 1360  TOOK ON 5  DUMPED 6
BUS 4 ARRIVED AT 1830  LEFT AT 1900  TOOK ON 7  DUMPED 7.
```

For this run, all passenger journeys were set to one stop only.

With the interval changed to 60 s for the buses and the mean time between passenger arrivals reduced to 10 s, bunching and overtaking of buses can be seen, as follows.

```
AT STOP 1
BUS 1 ARRIVED AT 30    LEFT AT 60    TOOK ON 3  DUMPED 0
BUS 2 ARRIVED AT 90    LEFT AT 120   TOOK ON 3  DUMPED 0
BUS 3 ARRIVED AT 150   LEFT AT 180   TOOK ON 3  DUMPED 0
BUS 4 ARRIVED AT 210   LEFT AT 240   TOOK ON 3  DUMPED 0
```

AT STOP 2
BUS 1 ARRIVED AT 90 LEFT AT 180 TOOK ON 9 DUMPED 3
BUS 2 ARRIVED AT 150 LEFT AT 165 TOOK ON 0 DUMPED 3
BUS 3 ARRIVED AT 210 LEFT AT 240 TOOK ON 3 DUMPED 3
BUS 4 ARRIVED AT 270 LEFT AT 300 TOOK ON 3 DUMPED 3

AT STOP 3
BUS 2 ARRIVED AT 195 LEFT AT 385 TOOK ON 19 DUMPED 0
BUS 1 ARRIVED AT 210 LEFT AT 255 TOOK ON 0 DUMPED 9
BUS 3 ARRIVED AT 270 LEFT AT 285 TOOK ON 0 DUMPED 3
BUS 4 ARRIVED AT 330 LEFT AT 345 TOOK ON 0 DUMPED 3

AT STOP 4
BUS 1 ARRIVED AT 285 LEFT AT 565 TOOK ON 28 DUMPED 0
BUS 3 ARRIVED AT 315 LEFT AT 315 TOOK ON 0 DUMPED 0
BUS 4 ARRIVED AT 375 LEFT AT 375 TOOK ON 0 DUMPED 0
BUS 2 ARRIVED AT 415 LEFT AT 510 TOOK ON 0 DUMPED 19.

In a fully stochastic version of the model, we would of course collect statistical summaries of the results of several runs.

10.9 FURTHER BATTLES

We have frequently mentioned the fact that many different models can be devised for the same physical situation. We now return to a topic previously considered in chapter 8 where we used a differential equation model to predict the outcome of a battle between two armies. To be more precise, this was a continuous deterministic model. No randomness was involved and each army was represented as a continuous function of time. In the model the progress of the battle was represented by the graph shown in Fig. 10.17.

Time is of course a truly continuous variable but the numbers of troops still fighting on both sides at time t are really integers. We now wish to study this problem using two other models, both of which are *discrete*, the first being deterministic and the second stochastic.

Discrete deterministic model

For this we represent our two armies by integer variables x and y and also let time progress in a series of discrete jumps. We recall that in section 8.5 we made the following explicit assumptions. In 1 h, each surviving X soldier kills 0.1 Y soldiers; in 1 h, each surviving Y soldier kills 0.15 X soldiers. We started with $x = 5000$ and $y = 10\,000$ and we found that the battle ended with $x = 7906$ and $y = 0$ after about 5.8 h. Let us choose 0.1 h as a time step and use x_n and y_n to denote the number of X and Y troops still alive after n time steps (where n is necessarily a positive integer). We shall take the above

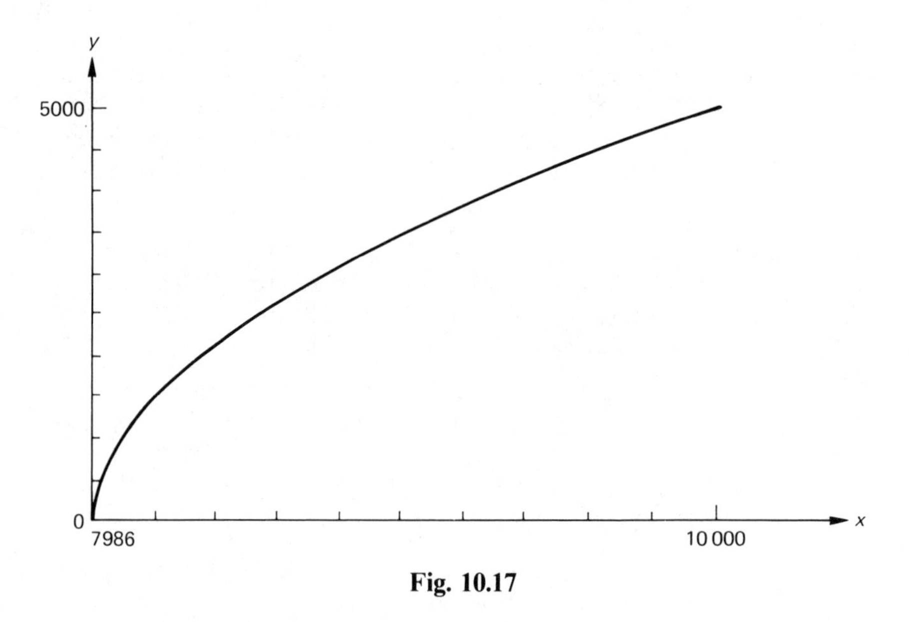

Fig. 10.17

assumptions to be equivalent to the following. In one time step, each X soldier kills 0.01 Y soldiers; in one time step, each Y soldier kills 0.015 X soldiers. The model can now be written as

$$\left\{\begin{matrix}\text{troops alive at}\\ \text{next time step}\end{matrix}\right\} = \left\{\begin{matrix}\text{troops alive at}\\ \text{this time step}\end{matrix}\right\} - \text{deaths},$$

i.e.

$$x_{n+1} = \text{INT}(x_n - 0.015y_n)$$

and

$$y_{n+1} = \text{INT}(y_n - 0.010x_n).$$

Note that, by using the INT function, we are making x_i and y_i into integer variables. We could have left x_i and y_i as discrete but *real* variables and ignored the decimal parts in our final evaluation. This leads to similar but slightly different results owing to the accumulated rounding errors. (Try it!)

Starting with $x = 10\,000$ and $y = 5000$, we can now very easily generate $x_1, y_1; x_2, y_2; \ldots$, from our two equations to give Table 10.16.

Discrete stochastic model

The fact that there are more troops on one side or that they are more effective fighters makes it more *likely* that the next single casualty in a battle will be from the other side. We can put together a model which allows chance to enter while at the same time taking account of this bias in the following way.

Table 10.16

n	x_n	y_n
0	10 000	5000
1	9 925	4900
⋮	⋮	⋮
57	7 856	71
58	7 854	−8

If at any stage in the battle, there are x and y troops surviving and fighting, then the probability that the next death is that of an X soldier can be represented by

$$\frac{0.15y}{0.15y + 0.1x}$$

and of a Y soldier

$$\frac{0.1x}{0.15y + 0.1x}.$$

Note that the two probabilities add up to 1 (the next death must be on one side or the other) and are in the correct ratio $bx:ay$.

Our stochastic model is therefore quite simple. We start with the values $x = 10\,000$ and $y = 5000$ and substitute in the above expressions to calculate the probabilities:

$$P(x \text{ becomes } x - 1) = 0.429,$$

$$P(y \text{ becomes } y - 1) = 0.571.$$

Taking a uniform [0, 1] random variable RND from a random-number generator, we let x become $x - 1$ if RND < 0.429; otherwise y becomes $y - 1$. We then *recalculate* the probabilities and take another random number, and so on. The graph describing the progress of the battle can be represented as in Fig. 10.18.

We start from the point $x = 10\,000$, $y = 5000$, and move in regular horizontal or vertical jumps along the grid until we finally reach one of the axes and one army or the other has been reduced to zero. Three particular runs from the same initial conditions gave the following results: $y = 0$ when $x = 7958$; $y = 0$ when $x = 7961$; $y = 0$ when $x = 7895$. By carrying out several runs, we can estimate the *spread* of likely results under identical conditions. Note that it *is* possible for army Y to win (but very unlikely).

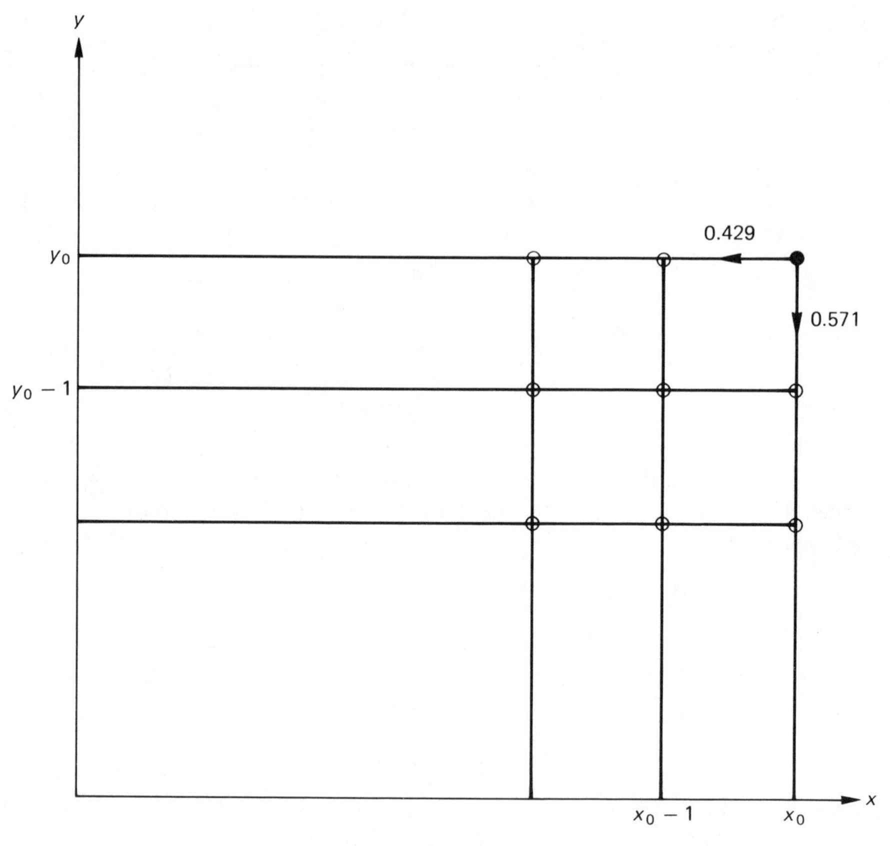

Fig. 10.18

Comparing the models

Why do the figures obtained from the deterministic discrete model not quite agree with those given by the continuous model? Note that in the discrete model the x and y values remain constant for 0.1 h at a time. The 'kill rates' of 0.015 and 0.01 per time unit which we have used are *not* precisely equivalent to the instantaneous rates 0.15 and 0.1 in the continuous model.

With the discrete stochastic model, different runs will of course produce different results but, if we take the *mean* of a large number of runs, then we would expect to reach very similar conclusions about the eventual outcome of the battle, as in the other two models. Note that one deficiency of the discrete stochastic model is that time does not explicitly appear, so that we have no way of estimating the duration of the battle. This *could* be done, however, by extending the model to make the deaths events in a 'pooled Poisson process'. At any stage in the battle the time between deaths is then a negative exponentially distributed random variable with mean $1/(bx + ay)$; so we can simulate it by taking

$$\Delta T = -\frac{\ln(\text{RND})}{bx + ay}.$$

Note that at the beginning of the battle, when x and y are large, ΔT is likely to be smaller, while towards the end of the battle we get longer time intervals between deaths, as we would expect.

10.10 SNOOKER

Context

Television exposure has made snooker very popular. The expertise of the game's best players continues to be impressive and they can usually be relied upon to pot the balls from all distances and angles. Advice to beginners is to practise hard and often, until you learn how to strike the cue ball so that after impact the coloured ball is sent off at just the correct angle and speed to enter a pocket. Can a mathematical model help to give advice to beginners on where to aim with the cue ball and consequently play better snooker?

The pockets are narrow, but there is a little margin of error which allows a ball to enter a pocket even if your aim is not quite perfect. Human error is always present; so an interesting question is how much deviation can be allowed on a shot while still allowing the coloured ball to enter the pocket. Advice can then be given to the player on where to aim with the cue ball and beginners can learn how to play better snooker.

Problem statement

To fix ideas, a particular problem will be considered. We shall suppose that only the *pink* and *black* balls remain to be potted and that they are on their marker spots on the table. Let us suppose that the white cue ball happens to be on the table centre spot. This is shown in Fig. 10.19.

The dimensions of a snooker table are required and the necessary data have been marked in the diagram. The problem that we wish to solve is quite simply stated as 'How should the *pink* be potted, followed by the *black* in the easiest possible manner?' This means that we can obviously concentrate first on potting the pink.

Formulate a mathematical model

There are two features to be considered by the player as he sets out to strike the cue ball: where to aim and how fast to hit the cue ball. He may also be skilful enough to put some 'side' spin on the cue ball so that a good position can then be obtained for potting the black in the next shot. However, the

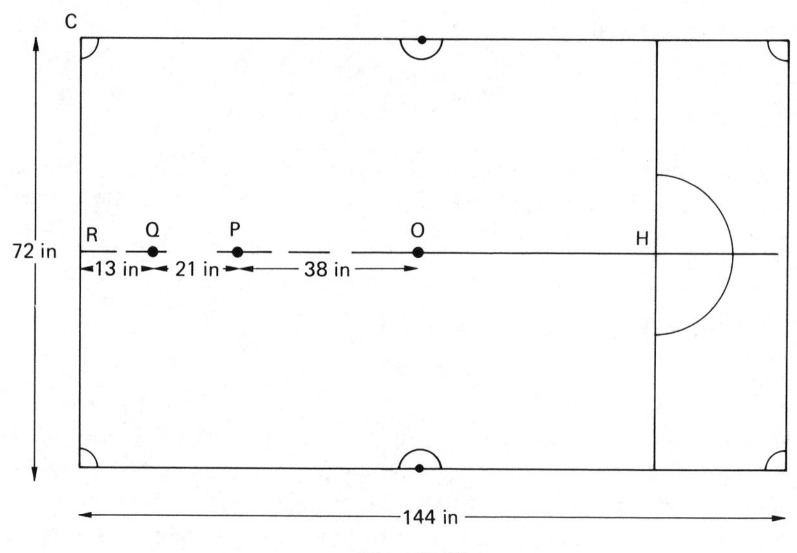

Fig. 10.19

investigation here will not take spin into account as the mathematics becomes a little difficult for an initial model.

The two issues of speed and direction can be separated in the first place. Looking at the directional problem, we quickly realise that this will involve geometrical work only—a mathematical model for the impact between snooker balls can be left until later. We assume that the pink ball moves off in the direction of the collision impulse of the two balls, and in turn this impulse will be taken always along the line joining the centres of the colliding balls, as they are assumed to be smooth. The problem is illustrated in Fig. 10.20.

From Fig. 10.20, it can be seen that the cue ball is aimed at an angle α to the initial line of centres and that the pink is then sent off towards the pocket C at an angle β. Our problem is to relate α to β, and to find how much

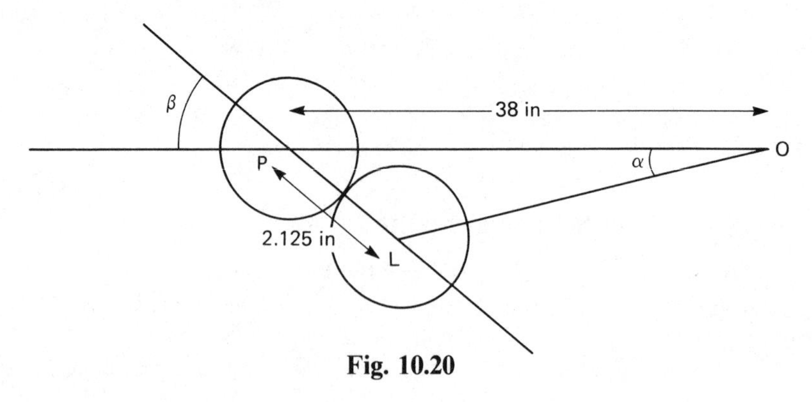

Fig. 10.20

Table 10.17

Description	Symbol	Units
Distances on a snooker table (see Fig. 10.20)	OP, PL, LO	in
Angles (see Fig. 10.20)	α, β	rad
Small variations in the angles	$\delta\alpha$, $\delta\beta$	rad
Initial speed of cue balls	u	in s^{-1}
Speed of pink ball	v	in s^{-1}
Velocity components of cue ball after impact	v_1, v_2	in s^{-1}
Elastic constant	e	—

variation is possible in making the shot so that the 'pot' is successful. We now list our variables in Table 10.17.

Obtain the mathematical solution

By using the 'sine' rule of trigonometry applied to the triangle OPL,

$$\frac{\text{OP}}{\sin(\alpha+\beta)} = \frac{\text{PL}}{\sin\alpha} = \frac{\text{LO}}{\sin\beta}. \tag{10.33}$$

A snooker ball has a diameter of 2.125 in; so, taking the data from Fig. 10.19, we can substitute into equation (10.33) to get

$$\frac{38}{\sin(\alpha+\beta)} = \frac{2.125}{\sin\alpha} = \frac{\text{LO}}{\sin\beta}.$$

Rearrangement of this gives, from the first pair of equations,

$$17.88\tan\alpha = \tan\alpha\cos\beta + \sin\beta. \tag{10.34}$$

Now β will be known from the table measurements as $\tan^{-1}(18/17)$ but, before using this value, we realise that it is the *perturbations* on α and β that are of interest. Denote a small change in α by $\delta\alpha$, and a similar small change in β by $\delta\beta$; then, by differentiating equation (10.34),

$$17.88\sec^2\alpha\,\delta\alpha = \sec^2\alpha\,\delta\alpha\cos\beta - \tan\alpha\sin\beta\,\delta\beta + \cos\beta\,\delta\beta$$

Substituting for α from equation (10.34), after a little manipulation we find that the formula relating the small changes in α and β is

$$\delta\alpha = \frac{(17.88\cos\beta - 1)\,\delta\beta}{320.7 - 35.76\cos\beta}.$$

Using $\beta = \tan^{-1}(18/17)$, this reduces to

$$\delta\alpha = 0.038\,078\,\delta\beta. \tag{10.35}$$

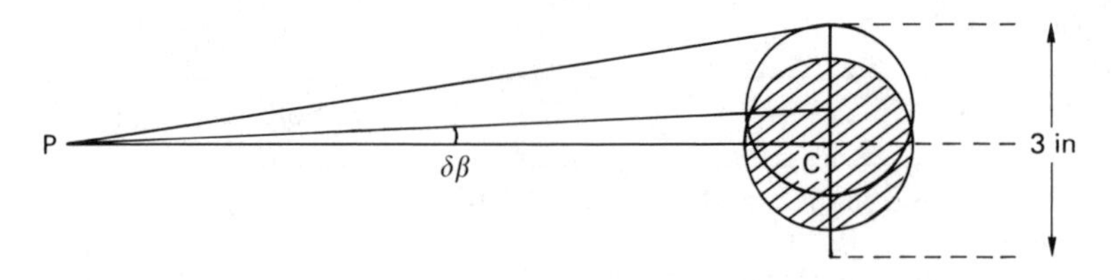

Fig. 10.21

This equation can be used to answer the first question on the accuracy of the shot. The entry to the pocket on a snooker table is of total width 3.0 in. Without discussing situations where the ball glances off the edge of the pocket and still drops in (notoriously unreliable!), this means that $\delta\beta$ can be calculated as shown in Fig. 10.21.

We find that $\delta\beta = 0.008\,829$ (measured in radians). Hence, using equation (10.35), $\delta\alpha = 0.000\,336$ rad.

This is a very tiny deviation. Its value can be expressed as a percentage change permitted in α, which itself is small. From equation (10.34),

$$\alpha = \tan^{-1}\left(\frac{\sin\beta}{17.88 - \cos\beta}\right)$$

$$= 0.042\,26 \text{ rad}$$

since $\beta = \tan^{-1}(18/17)$. The percentage deviation is $(\delta\alpha/\alpha) \times 100 = 0.8\%$.

Allowing for possible data inaccuracies and so on, we can say that the permitted deviation is about 1%; so the shot has to be struck very accurately.

Interpret the mathematical solution

From the point of view of the player, knowing the angle of impact and the error allowed does not necessarily help in lining up the shot. As the player bends across the table to strike the cue ball, it is the amount of obliqueness of the impact that has to be judged; it does not really help to know that $\alpha = 0.042\,26$ rad. To evaluate the obliqueness, we need to say how much of the pink should be visible from O as the balls collide. This is a slightly tricky point since the player sees the balls in line from a point some way back from O, and his eye will probably be above the plane of the table.

From Fig. 10.22 the angles γ and ρ are easily worked out by simple trigonometry to be $\gamma = 0.027\,96$ rad and $\rho = 0.029\,06$ rad. Hence, if we look at the pink from O, the percentage of the ball visible is

$$\frac{\alpha - \rho + \gamma}{2\gamma} \times 100 = \frac{0.041\,16 \times 100}{0.055\,92} = 73.6\%.$$

Thus, we have answered the next question, i.e. the fact that the obliqueness necessary is for about three-quarters of the pink to be visible from O.

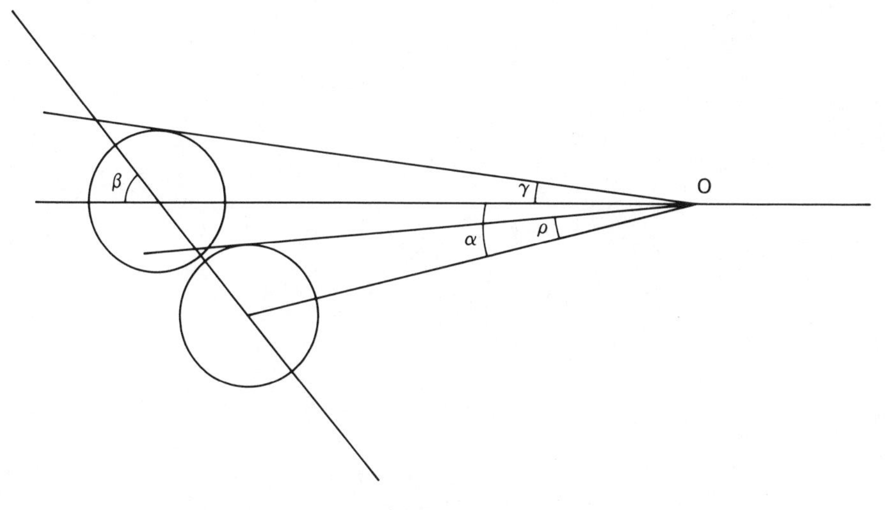

Fig. 10.22

We now turn our attention to the second part of the problem, namely the potting of the black ball. The path of the cue ball must be modelled first, which brings us to the question of how snooker balls behave on collision. For oblique impact, there are three properties of mechanical systems that can be applied.

(a) Total linear momentum is conserved along the line of centres of the two balls.
(b) The linear momentum of each ball is separately conserved in a direction perpendicular to the impulse.
(c) Newton's impact rule can be used; this states that the relative velocity after impact is proportional to the relative velocity before impact in a direction along the line of centres, where the constant of proportionality represents the elasticity of the materials in collision.

Application of (a)–(c) will enable the speed and direction of the cue ball after impact to be calculated. A diagram is necessary to help to follow the application (Fig. 10.23). In relation to the symbols marked in Fig. 10.19, we have, from (a),

$$u\cos(\alpha+\beta)=v_2+v, \tag{10.36}$$

from (b),

$$u\sin(\alpha+\beta)=v_1, \tag{10.37}$$

and, from (c),

$$v-v_2=(-e)(-u)\cos(\alpha+\beta). \tag{10.38}$$

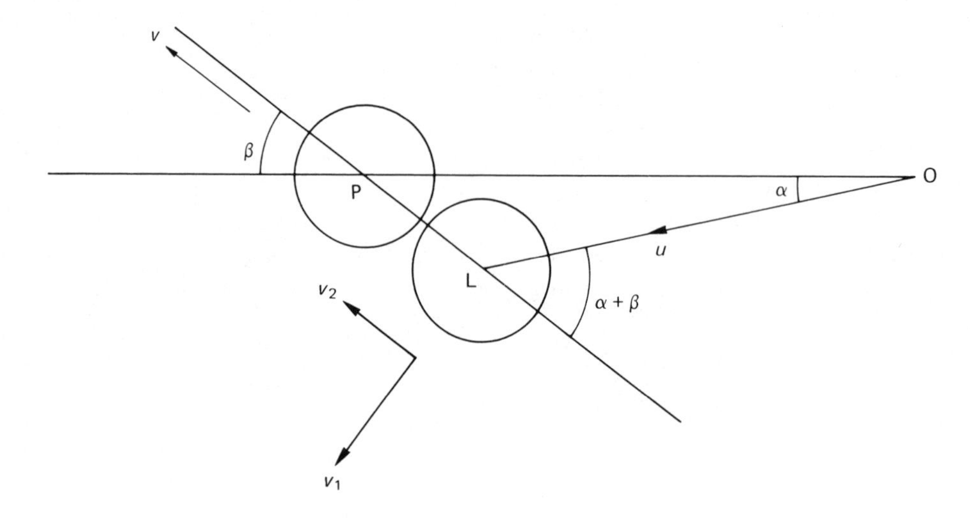

Fig. 10.23

A value is needed for the elastic constant *e*. It is possible that the cue ball will also strike the side cushion in its motion after colliding with the pink ball. Also, what will be a suitable value for the velocity of the cue ball just before impact? The following data are reasonably representative.

1 The elastic constant between two snooker balls equals 0.6.
2 The elastic constant between a ball and a cushion equals 0.8.
3 In motion along the table a snooker ball experiences a constant retardation of 0.7 ft s^{-2} (obtained by trial).
4 u is chosen so that the pink just drops in the pocket.

The motion of the cue ball can now be charted using the simple formula from particle kinematics which applies to motion with constant acceleration:

$$v^2 - u^2 = 2fs \tag{10.39}$$

(here u is the initial speed, v is the final speed, f is the acceleration and s is the distance travelled).

Now, as the pink just falls into the pocket, from equation (10.37) we get $v = 2 \times 0.7 \times 12\sqrt{36^2 + 34^2} = 28.8$ in s^{-1}. However, $\alpha = 2.42$ rad and $\beta = 46.64$ rad; so, from equation (10.36), $u = 55.02$ in s^{-1}.

The speed of the cue ball can now be calculated. From equations (10.36)–(10.38), $v_1 = 41.56$ in s^{-1} and $v_2 = 7.21$ in s^{-1}. These are velocity components with respect to the line PL. It is easier if the motion of the cue ball is related to OP, and a simple calculation gives the velocity component down the table as 35.17 in s^{-1} and that across as 23.29 in s^{-1} which on combining gives a velocity of 42.18 in s^{-1} at an angle of 33.51 rad with PR.

Now it is clear from the dimensions of the snooker table that the cue ball is heading for the back cushion. By using equation (10.37), we can calculate that the impact occurs with a speed of 32.88 in s^{-1}, at a distance of 20.05 in from the corner pocket. The cue ball then rebounds from this cushion with its velocity component normal to the cushion reduced by the factor 0.8 (specified earlier as the elastic constant between the ball and a cushion). Continuing in this way, the cue ball next rebounds off the side cushion and finally comes to rest at a point O′ 36.38 in from the end cushion and 15.95 in from the side. This is shown in Fig. 10.24.

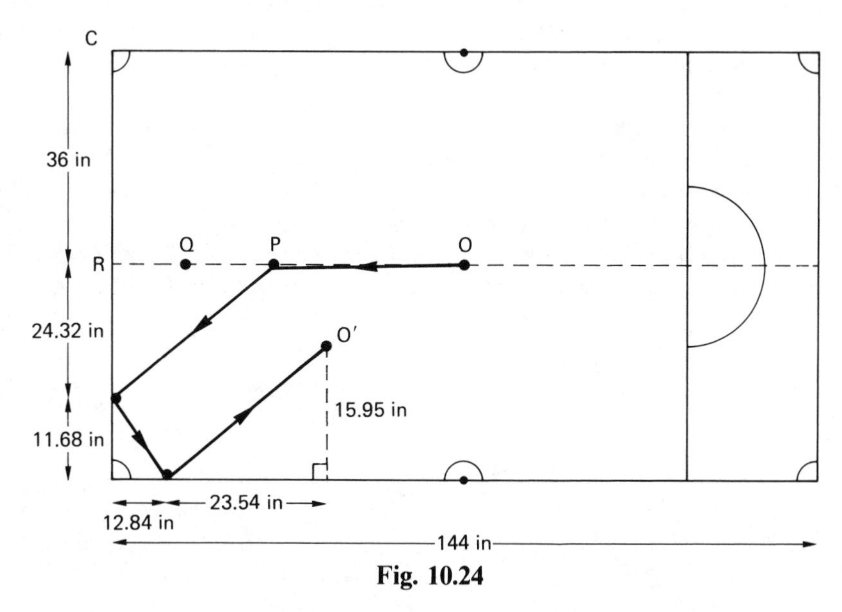

Fig. 10.24

Can the black ball now be potted, and if so what must be the angle of aim for the cue ball? Further calculation shows that we should aim so as to cover half the black ball on impact, with again a very small margin of error. Snooker is a difficult game to play well!

10.11 FURTHER MODELS

We conclude with a selection of modelling exercises for you to try your skills on. They are not arranged in any particular order and some may be easier than others.

Mileage

Employees of a firm send in claims for travelling expenses which are paid according to the mileage which they have travelled during the course of their work. The claim form requires each person to fill in details as follows.

	Journey		
Date	From	To	Total mileage
X	A	B	m

The clerk who processes the claims has been told to check the mileage figures using the map but she is very short of time and resorts to using a ruler and measuring the straight-line distance s from A to B. s is obviously an underestimate of m. Can you develop a mathematical model for estimating m from s: $m = f(s)$?

Check your model by measuring a number of distances on a map and give an overall assessment of the accuracy of your model.

Is it better to use one formula for smaller distances, say up to 50 miles, and a different formula for longer distances?

If the largest deviation d measured at right angles to AB is also measured can the value of d be incorporated into the model to improve the accuracy (Fig. 10.25)?

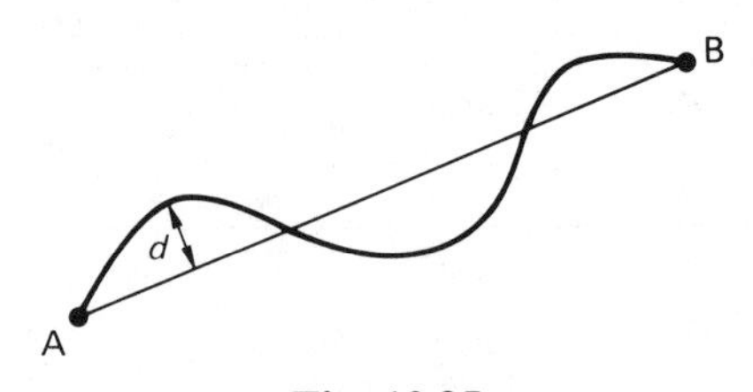

Fig. 10.25

Heads or tails

The result of tossing a coin is normally regarded as a random event with equal probabilities of heads and tails. In reality of course the outcome is entirely dependent on the initial conditions which set the coin in motion and, if these conditions and all the other relevant properties were known, we could predict the outcome.

Figure 10.26 illustrates a simple model in which the coin is represented by a straight-line segment of length l, held initially at an inclination θ to the horizontal and with its centre at height h above a horizontal table.

Suppose that the 'coin' is set spinning with angular speed $\mathrm{d}\theta/\mathrm{d}t$ and falls under gravity so that its centre moves along the vertical line shown in the sketch. Assume that the outcome is determined when one end or the other of the coin hits the table, i.e. do not try to model 'bouncing' off the table.

The objective is to discover how the outcome (heads or tails) depends on the values of h, l, θ and $\mathrm{d}\theta/\mathrm{d}t$.

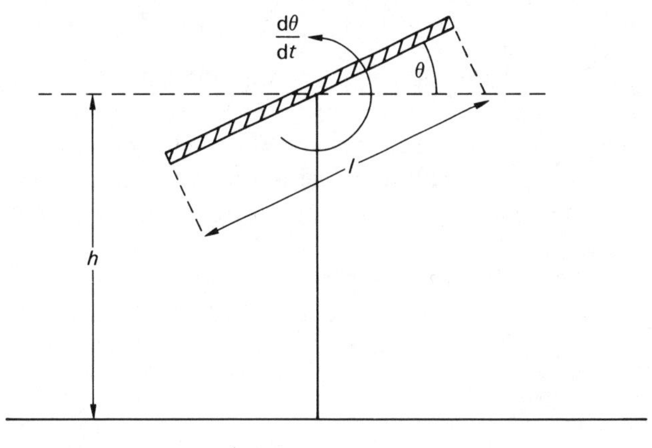

Fig. 10.26

Picture hanging

A picture (modelled as a rectangle) is to be suspended from a hook on a vertical wall by means of a piece of string attached to the back of the picture at two points as shown in Fig. 10.27.

The relevant parameters are the lengths a, b, c and d indicated in the sketch, the length l of the string and the coefficient of friction between the picture and the wall.

Develop a mathematical model which will enable you to find how the tension in the string depends on the other parameters, and in particular how to choose the length of the string and the points of attachment so that the string is least likely to break.

Also find the angle α of inclination.

For what range of values of the parameters will the string be visible above the top edge of the painting when viewed horizontally?

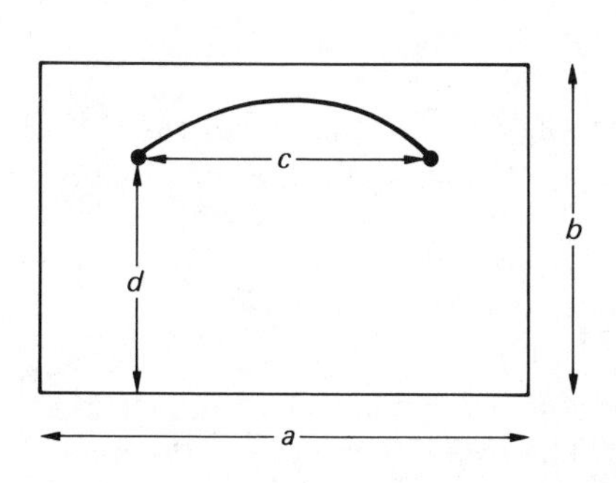

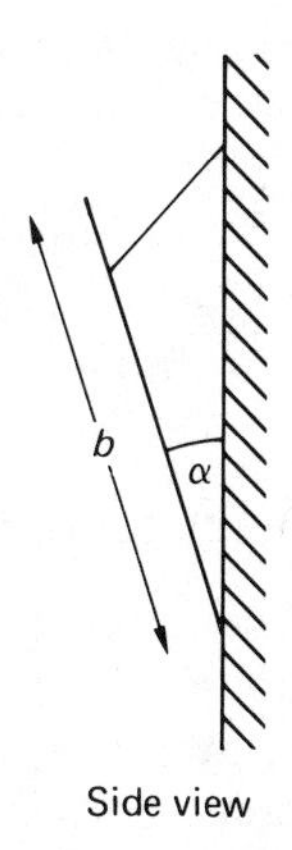

Fig. 10.27

Motorway

A motorway is to be built between cities A and B. City B is 20 km due south and 30 km due east of city A and there is a range of mountains running approximately east–west which lies between them. The cost of motorway building is related to the nature of the terrain and in Fig. 10.28 the whole area has been modelled in a very simple way using three different terrain types in parallel east–west strips.

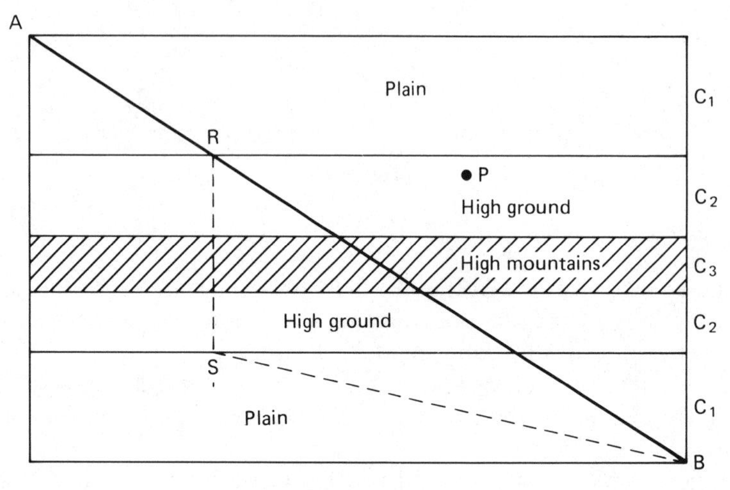

Fig. 10.28

Your task is to develop a mathematical model which will help to determine the cheapest route for the motorway given the relative costs per kilometre of construction for the three types of terrain and the widths of the strips. In the diagram the straight line route AB is obviously the shortest in distance but will not necessarily be the cheapest. A route such as ARSB has a short mountain section but is this the best solution? How would you adapt your model to incorporate the following two constraints.

1 When the route changes direction, the angle made must be at least 140°.
2 The route must pass through a given point such as P in order to merge with an existing road.

Vehicle-merging delay at a junction

Figure 10.29 shows a single-lane traffic stream on the major road and a minor approach road (no right turns allowed). The traffic flow rate on the major road is known, say q vehicles per hour. The time gaps (measured at a fixed point) between vehicles on the main road can be assumed to be independent random variables.

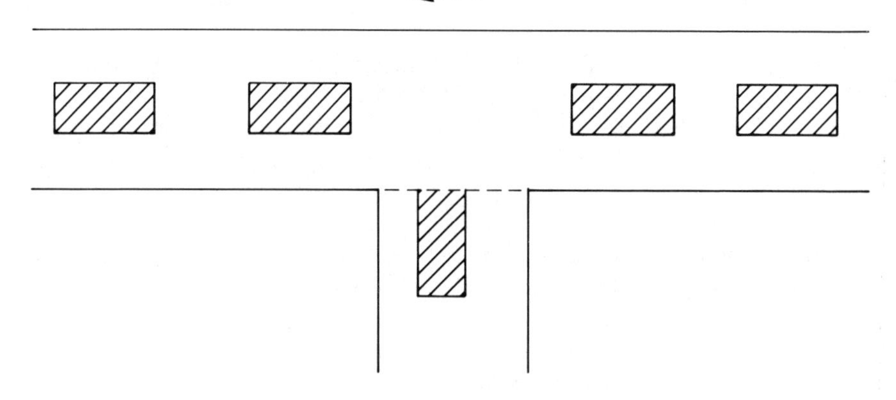

Fig. 10.29

Assume that vehicles arrive at random times on the minor road and that a driver will merge with the stream of traffic on the main road if he finds a time gap sufficiently large for his safe entry. We can model this by assuming the driver will always reject time gaps less than some minimum time gap T and that merging only takes place for time gaps greater than T. The delay time of the driver is zero if an acceptable time gap is available at the instant that he arrives at the junction; otherwise the driver has to wait for an acceptable gap.

Make what you think are reasonable assumptions about the values of q and T and any other parameters involved in the model. Find the distribution of delay times at the junction and investigate how this distribution depends on the other parameters.

Family names

You are asked to help with a sociological study on the survival and extinction of surnames.

Consider a closed community with N individuals at time $t = 0$ with K different surnames. Assume that all the children of any marriage take the father's surname.

Make other assumptions as necessary and set up a model with the aim of answering the following questions.

1 What will be the distribution of surnames after x generations?
2 What is the probability that a particular surname becomes extinct?
3 How long will a particular surname survive on average?

Estimating animal populations

One method of estimating the size of a living population occupying a finite region, e.g. fish in a lake, or squirrels in a wood, is to capture a few individuals and mark or tag them.

Suppose that x animals are caught, marked and then released. Some time later, n animals are caught and y of them are found to be marked. What is the best estimate that we can make of the total number N of animals living in the region? How accurate is this estimate likely to be and on what does the accuracy depend?

Derive 95% confidence limits for N.

Develop a simulation model to check your answers.

What advice would you give on the choice of values for x and n?

Simulation of population growth

Consider the life of an organism as the sequence birth–childhood–adulthood–old age–death. Suppose that an individual is capable of producing offspring only during adulthood and that the production of offspring can be modelled by a Poisson process. Make suitable assumptions about the duration of the three stages childhood, adulthood and old age.

Construct a simulation model which can be used to study the growth of a population starting from three young males and three young females at time $t=0$.

Find the size $N(t)$ of the population at a later time t and also at $t(n)$, the time that it takes for the population to grow to a size n. You should carry out a number of runs of the model and calculate mean values.

Needle crystals

In many chemical and metallurgical processes a surface film of needle-shaped crystals is formed as a substance cools. The crystals appear to start to grow from random points ('nuclei') and continue to grow in directly opposed directions. The position of a nucleus and the direction of growth appear to be totally random. The rate of growth stays reasonably constant. The crystals do not push each other apart when they come into contact; in fact, when a growing end touches another segment, that end stops growing and remains attached to the segment that blocked it (Fig. 10.30).

We need to know the mean number of crystal ends per unit area which are growing (unblocked) at time t. No method is known for finding this number exactly. You are asked to help as mathematical modellers.

1 Select appropriate variables and parameters for defining the problem.
2 Develop a mathematical model with associated software to find the number $n(t)$ of growing ends at time t.
3 Use your model to find how $n(t)$ depends on the other parameters in the problem.

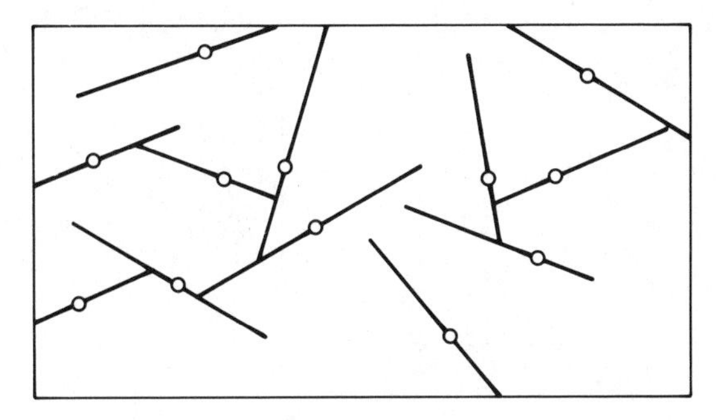

Fig. 10.30

Car parking

A training manual for car drivers wishes to offer the following advice for parking a car neatly into a space in a line of parked cars standing next to a kerb. For backing in, first overshoot the space by $x\%$ of a car length and stand off by $y\%$ of a car width (Fig. 10.31).

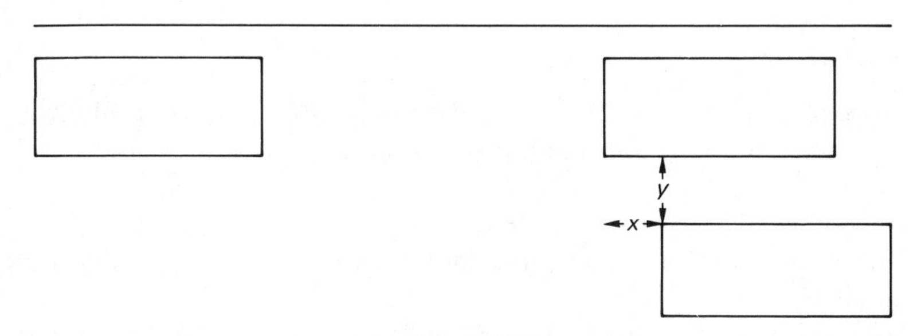

Fig. 10.31

1 You are asked to develop a mathematical model which will enable the author of the manual to substitute appropriate values for x and y.
2 What is the minimum gap length L in order for parking to be possible without driving over the kerb?
3 Find similar answers for the problem when driving into the space front wheels first.
4 After parking neatly and doing your shopping, you return to find that another driver has boxed you in. You may still be able to drive out by suitable manoeuvres. Use your mathematical model to investigate what manoeuvres may be necessary.

Overhead projector

Figure 10.32 shows a lecture room with 28 desks and an overhead projector and screen. The students at the sides near the front complain that they cannot

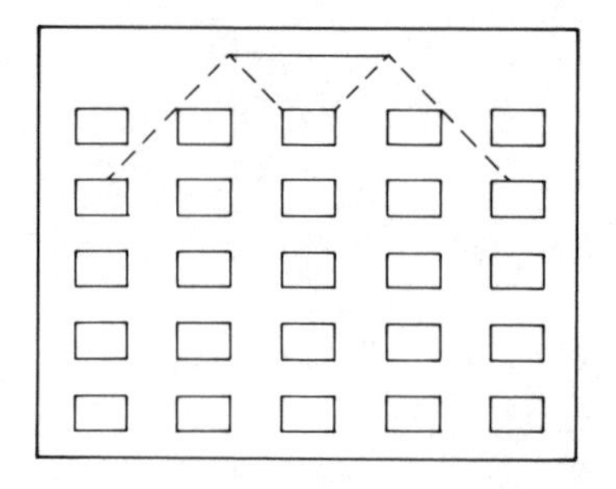

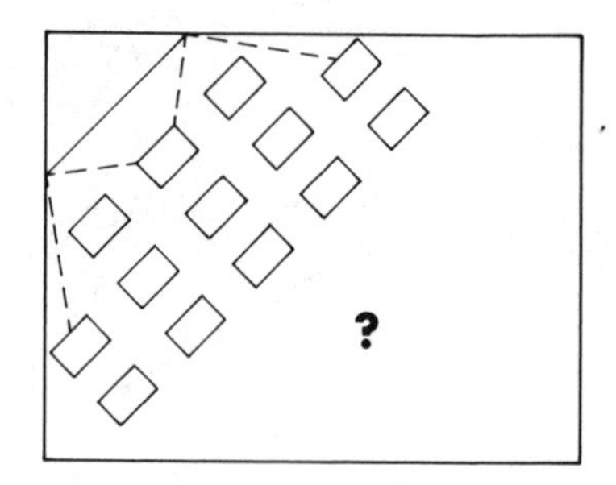

Fig. 10.32

see the screen. The lecturer thinks that it might be better to position the projector and screen in one corner of the room and to rearrange the desks.

Will this increase the number of students who can see the screen properly? Develop a general model which can be used to answer this question for a rectangular room of any dimensions.

Sheep farming

A prospective sheep farmer has x m^2 of land available for grazing and wishes to have answers to the following questions.

1 How many sheep should he keep?
2 How much of the summer grass should he store as feed for the winter months?
3 What proportion of female lambs should he retain each year for breeding?

You are asked to develop a mathematical model to help to answer these questions.

Background information

For a particular type of grass (Perennial Ryegrass) on a lowland site the following are approximations to the average growth rate.

	Winter	Spring	Summer	Autumn
Daily growth/g	0	3	7	4

In most common breeds of sheep the ewes produce one, two or three lambs each year until they are about 5–8 years old, when they are sold. If we assume that ewes are kept until they are 5 years old, then the following are the average numbers of lambs born per ewe in each age group.

Age/years	0–1	1–2	2–3	3–4	4–5
Lambs born	0	1.8	2.4	2.0	1.8

The following are average feeding requirements per animal during the year.

Daily requirements/kg of grass	Lambs	Ewes
Winter	0	2.10
Spring	1.00	2.40
Summer	1.65	1.15
Autumn	0	1.35

BIBLIOGRAPHY

There are several books dealing with mathematical modelling at an introductory level, of which the following are a selection.

Andrews, J. G., and Maclone, R. R., *Mathematical Modelling*, Butterworths, 1976.

Bender, E. A., *An Introduction to Mathematical Modelling*, Wiley, 1978.

Burghes, D., Huntley, I., and McDonald, J., *Applying Mathematics: A Course in Mathematical Modelling*, Ellis Horwood, 1982.

Burghes, D. N., and Borrie, M., *Modelling with Differential Equations*, Thornes, 1981.

Cross, M., and Moscardini, A., *Learning the Art of Mathematical Modelling*, Ellis Horwood, 1984.

Deo, N., *Systems Simulation with Digital Computer*, Prentice-Hall, 1983.

Dym, C. L., and Ivey, E. S., *Principles of Mathematical Modelling*, Academic Press, 1980.

Giordano, F. R., and Weir, M. D., *A First Course in Mathematical Modelling*, Wadsworth, 1985.

James D. J. G., and McDonald, J. J. (eds), *Case Studies in Mathematical Modelling*, Thornes, 1981.

Medley, D. G., *An Introduction to Mechanics and Modelling*, Heinemann, 1982.

Morgan, B. J. T., *Elements of Simulation*, Chapman and Hall, 1984.

Neelamkavil, F., *Computer Simulation and Modelling*, Wiley, 1987.

Saaty, T. L., and Alexander, J. M., *Thinking with Models*, Pergamon, 1981.

Townend, M. S., *Mathematics in Sport*, Ellis Horwood, 1984.

SOLUTIONS TO EXERCISES

CHAPTER 4

EXERCISES 4.2

1 1.13 DM l^{-1}.
2 1.79 m s^{-1}.
3 1 N $= 10^5$ dyn $= 7.23$ pdl.
1 pdl $= 1.38 \times 10^4$ dyn.
4 1 million ft^3 $day^{-1} = 0.3278$ m^3 s^{-1}.
5 Milk.
6 The track with a perimeter of 440 yd is longer by 2.34 m.
7 7.27×10^{-5} rad s^{-1}.

EXERCISES 4.4

1 N m^2 kg^{-2}.
2 $ML^{-7}T^{-2}$.
3 (a) Correct.
(b) Error.
(c) Correct.
(d) Error.
4 (a) Error.
(b) Correct.
(c) Error.
5 Incorrect.
6 $a = \frac{1}{2}$, $b = \frac{1}{2}$ and $c = 0$. Thus, $u = k\lambda^{1/2}g^{1/2}\rho^c$.

CHAPTER 5

EXERCISES 5.2

1 Temperature outside and temperature inside.
Cost.
Heat saving.
2 See chapter 3.
3 Height of thrower.
Angle of projection.
Speed of throw.
4 Number of lifts available.
Correct lift positions, number of other users, speed of operation of lifts.

EXERCISES 5.4

1 (a) (i) No. (ii) Yes. (iii) No. (iv) Yes.
(b) (i) Yes. (ii) No. (iii) No. (iv) No.
(c) (i) No. (ii) No. (iii) Yes. (iv) No.
(d) (i) No. (ii) No. (iii) Yes. (iv) No.
(e) (i) No. (ii) No. (iii) Yes. (iv) No.
(f) (i) Yes. (ii) No. (iii) No. (iv) No.
(g) (i) No. (ii) Yes. (iii) No. (iv) Yes.
(h) (i) No. (ii) Yes. (iii) No. (iv) No.
2 (a) (i) Yes. (ii) No. (iii) Yes.
(b) (i) No. (ii) Yes. (iii) No.
(c) (i) Yes. (ii) Yes. (iii) Yes.
3 (i) a gives a vertical shift.
b affects magnitude of expression.
c affects rate of decay.
(ii) a magnifies the expression.
b shifts maximum point.
c gives a vertical shift.

EXERCISES 5.5

1 (c) is correct.
2 (e) is correct.
3 kLU^2/D.
$[k] = ML^{-1}$.
kg m^{-1}.

EXERCISES 5.6

1 $A + B\sin(\pi t/12) + C\sin(\pi t/4380)$.
2 $A(x) = a/x$, $B(y) = b/y$, where $a/x + b/y = c$.
3 $A\sin^2(\pi t/6)\exp(-0.1t)$.

EXERCISES 5.7

1 (i) c is the smallest term, and $\frac{x}{b}$ the largest term.

(ii) ab is the smallest term, and $\frac{ax}{b}$ the largest term.

(iii) $\frac{c}{ax^2}$ is the smallest term, and $\frac{bx^3}{a}$ the largest term.

2 (i) $x^2 + a$.

(ii) $bx^2 + \frac{a}{x}$.

(iii) $a\sqrt{x}$.

3 (i) $0.01x^3$.

(ii) $\frac{0.2}{x} + 1$.

(iii) $\sqrt{0.001x}$.

4 (i) $0.001 + 0.0001x$.

(ii) x.

(iii) $\frac{0.1}{x^2} + \frac{0.1}{x}$.

CHAPTER 6

EXERCISES 6.3

1 $D = 0.275(Y - 1948) + 13.9$ which is valid over a restricted range of Y.
2 26 miles 385 yd is 45 413.77 m.
The result for 1968 could be because the Olympic Games were held at a high altitude.
$T = 2.00 + a\exp[-b(Y - 1948)]$.
3 Price $\approx \frac{5}{6} \times$ mass.
5 (a) $T^2 = kR^3$. Confirm from logarithmic plot.
(b) $R = 14.96\,(4 + 3 \times 2^{n-2})$, where $n = 2, 3, \ldots, 8$ (the Titus–Bode Law).

EXERCISES 6.5

1 $x > 1.472$.
2 0.314 m^2.
3 Approximately 50 ft.

EXERCISE 6.6

1 Using the least squares criterion, the second model is more accurate.

CHAPTER 8

EXERCISES 8.2

1 $\dot{N} = 0.02 \sin^2(\pi t/4)N - 0.1$, $N(0) = 5$. t measured in years, N in thousands.
2 $T(t) = 18 + 42 \exp(-0.0906t)$. Cools to 30 °C in 13.82 min. $T(10) = 34.97°$.
3 $V(t) = (1.279 - 10.0399t)^3$. V is in m^3, t is in h.
4 $Y = \alpha^2/4\beta$.

EXERCISES 8.3

1 (a) False.
(b) False.
(c) False.
(d) False.
(e) (i) No. (ii) Velocity zero. (iii) No.
(f) (i) No. (ii) Yes. (iii) Yes. (iv) No.

2 (a) Moment of applied force about hinge axis is required. This is the product of force and perpendicular distance. So if the distance is reduced then the force must be increased for the same effect.
(b) Friction opposes motion and is proportional to the normal reaction between the object and the ground. This reaction is larger when pushing.
(c) Circular motion has force towards the centre which increases as the square of the angular velocity.
(d) As capsule moves in an approximately circular orbit, the astronaut experiences a balance between gravitation and centrifugal force.

3 The differential equation has the form

$$m\ddot{x} + k\dot{x} + \lambda x = \lambda y(nt) + kny'(nt)$$

where x is the vertical car body displacement, y is the ramp profile, λ is the spring constant, k is the damper constant, n is the speed of the car at time t after the ramp is hit and m is the mass of the car.

4 For case A, the motion can be described by the equations

$$Tr = -I\ddot{\theta}, \qquad T\cos(\theta - \phi) = G\phi$$

and

$$\frac{T}{\sin\phi} = \frac{b}{\sin(\theta - \phi)} = \frac{l}{\sin\theta}$$

where T is the force between the arm and the door, G is the spring force, $r = \text{PB}$, $b = \text{BS}$, $l = \text{PS}$, θ is the angle at which the door is open, $\phi = \text{B}\hat{\text{S}}\text{P}$ and I is the moment of inertia of the door.

[case B—no solution supplied]

5 Variable mass problem. The equations of motion are

$$pgx - T = \frac{\mathrm{d}}{\mathrm{d}t}(pxv)$$

and

$$T - \mu g(l - x) = \frac{\mathrm{d}}{\mathrm{d}t}\{\rho(l - x)v\}$$

where x is length of vertical chain, l the total chain length, T the tension in the chain at the deck edge, ρ the mass per unit length of the chain, μ the coefficient of friction and v the velocity.

EXERCISES 8.4

1 Equations of motion are

horizontally:

$$\ddot{x} + k\dot{x} + KV = 0; \qquad x(0) = 0; \qquad \dot{x}(0) = u\cos\alpha + \bar{u}$$

vertically: $\ddot{y} + k\dot{y} + g = 0; \qquad y(0) = h; \qquad \dot{y}(0) = u\sin\alpha$

where V is the wind speed, $\bar{u}$ the run-up speed and u the throw speed.

2 Conditions for the ball to enter the basket:

$$3.05 - h = 4.6\tan\alpha - \left(\frac{g}{2}\right)(4.6)^2\frac{\sec^2\alpha}{n^2}; \qquad \phi \geqslant \sin^{-1}\left(\frac{5}{9}\right)$$

where ϕ is the angle of path with the horizontal on entry, u the initial throw speed and angle α, and h the height of the thrower.

EXERCISES 8.5

1 Approximately 53 h.
2 $\mathrm{d}x/\mathrm{d}t = -ay + R_1$, $\mathrm{d}y/\mathrm{d}t = -bx + R_2$.
3 $\dot{X} = aXF - bX$
$\dot{Y} = cYF - dY$
$\dot{F} = -eX - fY$

INDEX